FRESH WATER POLLUTION. II:
Radioactive Pollutants

Papers by
Forrest E. Knowles, Jr., H.L. Krieger, Edmond J. Baratta, Chester L. Nayfield, D.S. Woodhead, A.A. Cigna, Robert N. Snelling et al.

MSS Information Corporation
655 Madison Avenue, New York, N.Y. 10021

Library of Congress Cataloging in Publication Data

Main entry under title:

Radioactive pollutants.

(Fresh water pollution, 2)
1. Radioactive pollution of water. I. Knowles, Forrest E. II. Series. [DNLM: 1. Water pollution--Collected works. 2. Water pollution, Radioactive--Collected works. WA 689 F887]
TD420.F75 vol. 2 [TD427.R3] 628.1'68'08s
ISBN 0-8422-7090-6 [628.1'683] 72-13707

TABLE OF CONTENTS

CREDITS AND ACKNOWLEDGEMENTS

Baratta, Edmond J.; and Forrest E. Knowles, "Statistical Limits of Accuracy and Precision of Gross Alpha and Beta Radioactivity Measurements in Water," *Radiological Health Data and Reports*, 1971, 12:195-198.

Bureau of Radiological Health, State of California, Department of Public Health, "Radioactivity in California Waters, January-December 1969," *Radiological Health Data and Reports*, 1971, 12:513-518.

Bureau of Radiological Pollution Control, New York State Department of Environmental Conservation, "Radioactivity in New York Surface Water, July-December 1969 and January-June 1970," *Radiological Health Data and Reports*, 1971, 12:473-477.

Cigna, A.A.; G.F. Clemente; and F.G. Giorcelli, "On ^{134}Cs in Rainwater from 1960 to 1969," *Health Physics*, 1971, 21:667-672.

"Environmental Levels of Radioactivity at Atomic Energy Commission Installations. I. Feed Materials Production Center (July-December 1970). II. Portsmouth Area Gaseous Diffusion Plant (July-December 1970)," *Radiological Health Data and Reports*, 1971, 12:486-493.

Environmental Levels of Radioactivity at Atomic Energy Commission Installations. I. Atomics International (January-June 1970). II. Neutron Devices Department (January-June 1970)," *Radiological Health Data and Reports*, 1971, 12:113-117.

"Environmental Levels of Radioactivity at Atomic Energy Commission Installations. I. Lawrence Radiation Laboratory (July-December 1970)," *Radiological Health Data and Reports*, 1971, 12:529-538.

"Environmental Levels of Radioactivity at Atomic Energy Commission Installations. I. Oak Ridge Area (July-December 1969). II. Savannah River Plant (July-December 1970)," *Radiological Health Data and Reports*, 1971, 12:427-438.

Health and Safety Laboratory, U.S. Atomic Energy Commission, "Radiostrontium in Tap Water, July-December 1970," *Radiological Health Data and Reports*, 1971, 12:519.

Knowles, Forrest E., Jr.; and Edmond J. Baratta, "The Determination of Accuracy and Precision Limits for Tritium in Water," *Radiological Health Data and Reports*, 1971, 12:405-409.

Krieger, H.L.; and G.W. Frishkorn, "Evaluation of Ion-Exchange Surveillance Sampler for Analyzing Radioactive Liquid Effluents," *Health Physics*, 1971, 21:591-595.

Nayfield, Chester L.; Wallace Johnson; and Benjamin P. Prewitt, "Preoperational Radiological Surveillance of the Florida Power Corporation's Crystal River Power Reactor Site," *Radiological Health Data and Reports*, 1971, 12:441-449.

Office of Radiation Programs, Environmental Protection Agency, "Tritium Surveillance System, January-March 1971," *Radiological Health Data and Reports*, 1971, 12:384-387.

Office of Water Programs, Environmental Protection Agency, "Gross Radioactivity in Surface Waters of the United States, January 1971," *Radiological Health Data and Reports*, 1971, 12:472-473.

Office of Water Programs, Environmental Protection Agency, "Gross Radioactivity in Surface Waters of the United States, February 1971," *Radiological Health Data and Reports*, 1971, 12:512.

Snelling, Robert N., "Environmental Survey of Uranium Mill Tailings Pile, Mexican Hat, Utah," *Radiological Health Data and Reports*, 1971, 12:17-28.

"Water," *Radiological Health Data and Reports*, 1971, 12:421.

Woodhead, D.S., "The Biological Effects of Radioactive Waste," *Proceedings of the Royal Society of London: Biology*, 1971, 177:423-437. Reproduced by permission of the Controller of Her Britannic Majesty's Stationery Office.

PREFACE

This three-volume collection on fresh water pollution is an addition to the MSS Topics in Ecology series.

Papers published from 1971-1972 are presented providing current information on new developments in major aspects of fresh water pollution including bacteriological, chemical, thermal and radioactive pollution. Application of the latest technical devices for the control of water pollution is also discussed.

Radioactive Pollutants

WATER

The Environmental Protection Agency and other Federal, State, and local agencies operate extensive water quality sampling and analysis programs for surface, ground, and treated water. Most of these programs include determinations of gross beta and gross alpha radioactivity and specific radionuclides.

Although the determination of the total radionuclide intake from all sources is of primary importance, a measure of the public health importance of radioactivity levels in water can be obtained by comparison of the observed values with the Public Health Service Drinking Water Standards (*1*). These standards, based on consideration of Federal Radiation Council (FRC) recommendations (2–4), set the limits for approval of a drinking water supply containing radium-226 and strontium-90 at 3 pCi/liter and 10 pCi/liter, respectively. Higher concentrations may be acceptable if the total intake of radioactivity from all sources remains within the guides recommended by FRC for control action. In the known absence[1] of strontium-90 and alpha-particle emitters, the limit is 1,000 pCi/liter gross beta radioactivity, except when additional analysis indicates that concentrations of radionuclides are not likely to cause exposures greater than the limits indicated by the Radiation Protection Guides. Surveillance data from a number of Federal and State programs are published periodically to show current and long-range trends. Water sampling activities reported in *Radiological Health Data and Reports* are listed below.

[1] Absence is taken to mean a negligibly small fraction of the specific limits of 3 pCi/liter and 10 pCi/liter for unidentified alpha-particle emitters and strontium-90, respectively.

Water sampling program	Period reported	Issue
California	July–December 1968	August 1970
Interstate Carrier Drinking Water	1970	July 1971
Kansas	January–December 1969	September 1970
Minnesota	July–December 1969	May 1971
North Carolina	January–December 1967	May 1969
New York	January–June 1969	June 1970
Radiostrontium in Tap Water, *HASL*	January–December 1969	July 1970
Tritium in Community Water Supplies	1969	December 1970
Tritium Surveillance System	January–March 1971	July 1971
Washington	July 1968–June 1969	February 1971

REFERENCES

(*1*) U.S. PUBLIC HEALTH SERVICE. Drinking water standards, revised 1962, PHS Publication No. 956. Superintendent of Documents, U.S. Government Printing Office, Washington, D.C. 20402 (March 1963).

(*2*) FEDERAL RADIATION COUNCIL. Radiation Protection Guidance for Federal Agencies. Memorandum for the President, September 1961. Reprint from the Federal Register of September 26, 1961.

(*3*) FEDERAL RADIATION COUNCIL. Background material for the development of Radiation Protection Standards, Report No. 1. Superintendent of Documents, U.S. Government Printing Office, Washington, D.C. 20402 (May 1960).

(*4*) FEDERAL RADIATION COUNCIL. Background material for the development of Radiation Protection Standards, Report No. 2. Superintendent of Documents, U.S. Government Printing Office, Washington, D.C. 20402 (September 1961).

Tritium Surveillance System, January-March 1971

Office of Radiation Programs
Environmental Protection Agency

The Tritium Surveillance System is an expansion of previous tritium surveillance activities conducted by the Office of Radiation Programs, Environmental Protection Agency (EPA). The principal effort in the past by the Office of Radiation Programs related to tritium releases has been the Tritium in Surface Water Network. This network was established in 1964, to measure and monitor tritium concentrations in major river systems in the United States and to provide surveillance at surface water stations downstream from selected nuclear facilities. The network consisted of selected stations from existing water pollution sampling stations operated by the Office of Water Programs of EPA. The final data from this network for January–June 1970 were published previously (*1*).

Another effort of the Office of Radiation Programs was a tritium in precipitation program. This project was established in 1967 at selected Radiation Alert Network (RAN) stations covering the United States, including Alaska and Hawaii. The RAN is operated by the Office of Air Programs of EPA. The data from this project for July–December 1969, have been previously published (*2*). Due to the increased interest in tritium releases from nuclear facilities and the potential long-term accumulation in the environment, a national system was established to incorporate these projects or networks into one overall system.

Present network

The Tritium Surveillance System consists of 70 quarterly drinking water samples at the RAN stations, precipitation samples collected daily and analyzed monthly from 8 of the RAN stations (figure 1), and quarterly samples at 38 surface water stations (figure 2). The specific locations for the surface water sampling system were determined by examining the water drainage areas to assure that a representative sample from a large area or region was obtained and, if possible, incorporating several nuclear facility sites. All nuclear facilities that were operating, being constructed, or planned through 1975 were considered. Consideration was also given to the current surveillance programs of the States that will be involved in the collection of the samples. The

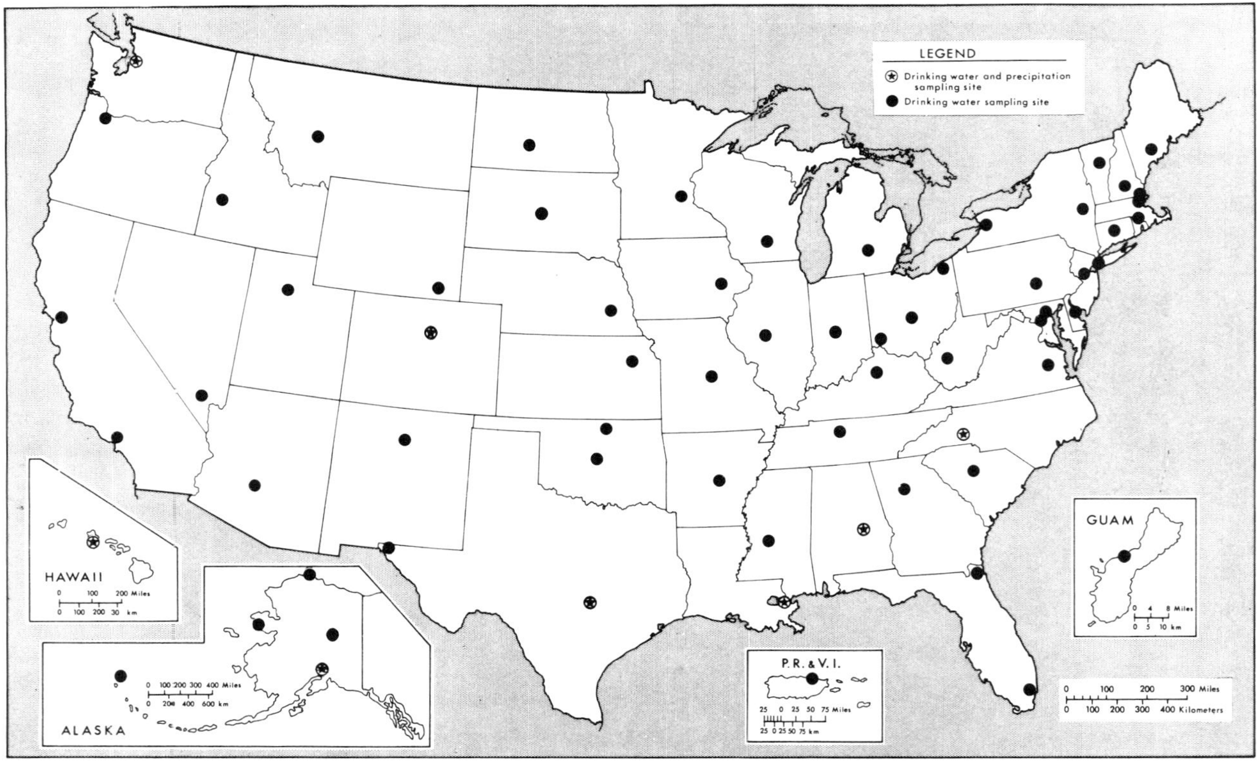

Figure 1. Drinking water and precipitation sampling locations for tritium surveillance system

Table 1. Tritium concentration in tap water (RAN stations) January–March 1971

Location		Date collected (1971)	Tritium concentration (nCi/liter ±2σ)
Ala:	Montgomery	2/1	[a]0
Alaska:	Anchorage	3/1	.6±.4
	Attu Island	1/7	0
	Fairbanks	1/7	0
	Juneau	1/26	0
	Kodiak	1/5	.4±.4
	Nome	1/7	0
	Point Barrow	1/14	1.2±.4
Ariz:	Phoenix	NS	
Ark:	Little Rock	1/26	[a]0
Calif:	Berkeley	1/5	0
	Los Angeles	1/4	0
C.Z:	Ancon	2/16	[a]0
Colo:	Denver	2/11	1.0±.4
Conn:	Hartford	1/11	.5±.4
Del:	Dover	1/11	0
D.C:	Washington	1/5	[a]0
Fla:	Jacksonville	1/22	[a]0
	Miami	1/25	[a]0
Ga:	Atlanta	2/1	[a]0
Guam:	Agana	1/4	0
Hawaii:	Honolulu	1/11	0
Idaho:	Boise	1/7	0
Ill:	Springfield	1/10	0
Ind:	Indianapolis	1/14	0
Iowa:	Iowa City	1/25	.9±.4
Kans:	Topeka	1/13	0
Ky:	Frankfort	1/25	[a]0
La:	New Orleans	1/22	[a]0
Maine:	Augusta	1/5	0
Md:	Baltimore	1/25	[a]0
Mass:	Lawrence	1/5	0
	Winchester	1/5	0
Mich:	Lansing	2/5	0
Minn:	Minneapolis	1/5	1.0±.4
Miss:	Jackson	1/25	[a]0
Mo:	Jefferson City	1/5	0
Mont:	Helena	1/5	1.0±.4
Nebr:	Lincoln	1/6	0
Nev:	Las Vegas	1/6	0
N.H:	Concord	1/11	.5±.4
N.J:	Trenton	1/11	0
N. Mex:	Sante Fe	NS	
N.Y:	Albany	1/1	0
	Buffalo	1/11	0
	New York City	1/29	0
N.C:	Gastonia	1/26	[a]0
N. Dak:	Bismarck	2/8	1.1±.4
Ohio:	Cincinnati	1/22	0
	Columbus	1/5	0
	Painesville	1/5	0
Okla:	Oklahoma City	1/25	[a]0
	Ponca City	1/25	[a]0
Oreg:	Portland	NS	
Pa:	Harrisburg	1/5	0
P.R:	San Juan	2/18	[a]0
R.I:	Providence	1/6	0
S.C:	Columbia	1/8	[a]0
S. Dak:	Pierre	1/7	1.3±.4
Tenn:	Nashville	1/26	[a]0
Tex:	Austin	1/27	[a]0
	El Paso	1/26	[a]0
Utah:	Salt Lake City	1/4	.5±.4
Vt:	Barre	1/4	0
Va:	Richmond	1/31	[a]0
Wash:	Seattle	NS	
	Spokane	1/8	0
W. Va:	Charleston	1/28	[a]0
Wisc:	Madison	1/21	0
Wyo:	Cheyenne	1/4	1.7±.4
Network average			0.2

[a] The minimum detection limit for this sample was 0.2 nCi/liter. All values equal to or less than 0.2 nCi/liter have been reported as zero. For all other zero numbers and for the calculation of the network average, the minimum detection limit was 0.4 nCi/liter and all values equal to or less than 0.4 nCi/liter have been reported as zero.

NS, no sample.

surface water samples are collected quarterly, either downstream from a nuclear facility or at a background station.

The tap water samples are collected by the RAN operators on a quarterly basis. The precipitation samples are also collected by the RAN operators on a daily basis.

All samples are sent to one of the EPA laboratories for analysis. Due to the different sensitivity of procedures used by the three laboratories, the minimum detectable concentration varies. Analytical values which are not statistically significant at the 2-sigma confidence level have been reported as zero.

Results and discussion

Table 1 presents the tritium concentrations in drinking water at the RAN stations for January–March 1971. The average tritium concentration was 0.2 nCi/liter.

The radiation dose resulting from the observed tritium concentrations in the drinking water may be evaluated by using the relationship derived by Moghissi and Porter (*3*) from recommendations of the International Commission of Radiological Units and Measurements (*4*). Assuming that the concentration of tritium in all water taken into the body is equal to that found in the drinking water and also that the specific activity of tritium in the body is essentially the same as that in the drinking water, then the radiation dose may be estimated.

The highest individual concentration of tritium observed in the drinking water was 1.7 nCi/liter during the first quarter. This corresponds to a dose of 0.3 mrem/a, or less than 0.2 percent of the Federal Radiation Council's Radiation Protection Guide (170 mrem/a) for an average dose to a suitable sample of the exposed population.

The tritium concentrations for the surface water samples are given in table 2. The highest tritium concentration was 3.7 nCi/liter for the

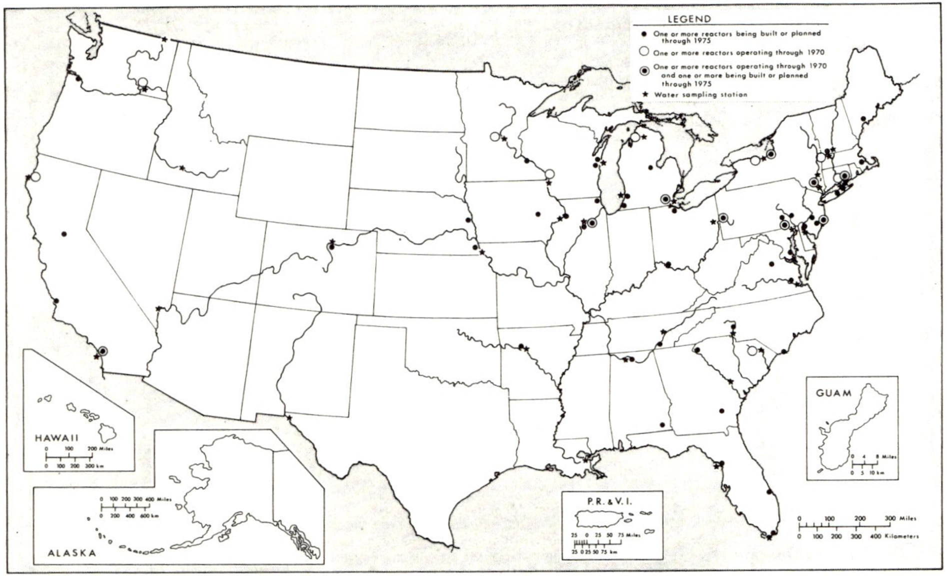

Figure 2. Surface water sampling locations for tritium surveillance system

Table 2. Tritium concentration in surface water, January–March 1971

Location		Water source	Facility	Collection date (1971)	Concentration (nCi/liter $\pm 2\sigma$)
Ala:	Decatur	Tennessee River	Browns Ferry	1/14	[a]0
Ark:	Morrilton	Arkansas River	Arkansas Nuclear	1/14	[a]0
Calif:	Eureka	Humboldt Bay	Humboldt Bay	1/28	0
	San Onofre	Pacific Ocean	San Onofre	3/4	0
Colo:	Greely	South Platte River	Fort St. Vrain	2/12	1.0±.4
Conn:	East Haddam	Connecticut River	Connecticut Yankee & Vermont Yankee	1/28	.6±.4
	Waterford	Long Island Sound	Millstone	1/28	.5±.4
Fla:	Crystal River	Gulf of Mexico	Crystal River	1/12	[a]0
	Key Largo	Biscayne Bay	Turkey Point	1/17	.3±.2
Idaho:	Buhl	Snake River	National Reactor Testing Station	1/7	0
Ill:	Moline	Mississippi River	Quad-Cities	1/29	0
	Morris	Illinois River	Dresden and Argonne	1/19	0
La:	New Orleans	Mississippi River	(Several)	1/14	[a]0
Md:	Conowingo	Susquehanna River	Peach Bottom and Three Mile Island	2/17	0
	Lusby	Chesapeake Bay	Calvert Cliffs	NS	
Mass:	Rowe	Deerfield River	Yankee	1/25	2.9±.4
Mich:	Charlevoix	Lake Michigan	Big Rock Point	3/21	0
	Monroe	Lake Erie	Enrico Fermi	3/15	0
	South Haven	Lake Michigan	Palisades	3/22	0
Minn:	Monticello	Mississippi River	Monticello	1/5	1.0±.4
Nebr:	Rulo	Missouri River	Fort Calhoun and Cooper	1/20	1.3±.4
Nev:	Boulder City	Colorado River	Background	NS	
N.J:	Bayside	Delaware River	Salem	NS	
N.Y:	Ossining	Hudson River	Indian Point	1/18	0
	Oswego	Lake Ontario	Nine Mile Point and R. E. Ginna	3/12	0
	Poughkeepsie	Hudson River	Background	1/13	0
N.C:	Charlotte	Catawba River	Wm. B. McGuire	1/14	[a]0
S.C:	Allendale	Savannah River	Savannah River Plant and Oconee	1/27	2.4±.2
	Hartsville	Lake Robinson	H. B. Robinson	3/10	.3±.2
Tenn:	Kingston	Clinch River	Oak Ridge	1/12	3.7±.2
Tex:	El Paso	Rio Grande	Los Alamos	1/11	.4±.4
Vt:	Vernon	Connecticut River	Vermont Yankee	1/4	.6±.4
Va:	Surry	James River	Surry	1/22	[a]0
Wash:	Northport	Columbia River	Background	2/10	.8±.4
	Pasco	Columbia River	Hanford	1/15	.9±.4
W.Va:	Wheeling	Ohio River	Shippingport	1/7	.5±.4
Wisc:	Two Creeks	Lake Michigan	Point Beach and Kewaunee	1/15	0
	Victory	Mississippi River	LaCrosse and Prairie Island	3/11	0
Network average					0.5

[a] The minimum detection limit for this sample was 0.2 nCi/liter. All values equal to or less than 0.2 nCi/liter have been reported as zero. For all other zero numbers and for the calculation of the network average, the minimum detection limit was 0.4 nCi/liter and all values equal to or less than 0.4 nCi/liter have been reported as zero.

NS, no sample.

first quarter. Assuming that the specific activity of tritium in the body is essentially the same as that in surface water, this concentration corresponds to a dose of 0.6 mrem/a, or 0.4 percent of the Radiation Protection Guide.

The monthly analyses for tritium in precipitation samples at the eight RAN stations are shown in table 3.

Table 3. Tritium concentrations in precipitation from RAN stations, January–March 1971

Location	Tritium concentration[a] (nCi/liter ± 2σ)		
	January	February	March
Alaska: Anchorage	NS	0.4±0.2	0.5±0.2
Ala: Montgomery	0	0	0
Colo: Denver	0	.5± .2	.8± .2
Hawaii: Honolulu	0	0	.2± .2
La: New Orleans	0	0	0
N.C: Gastonia	0	0	0
Tex: Austin	NS	NS	NS
Wash: Seattle	0	0	.3± .2

[a] The minimum detection limit for these samples was 0.2 nCi/liter. All values equal to or less than 0.2 nCi/liter have been reported as zero.
NS, no sample.

REFERENCES

(*1*) BUREAU OF RADIOLOGICAL HEALTH. Tritium in surface water network, January–June 1970. Radiol Health Data Rep 11:638–639 (November 1970).

(*2*) BUREAU OF RADIOLOGICAL HEALTH. Tritium in precipitation (RAN), July–December 1969. Radiol Health Data Rep 11:354–355 (July 1970).

(*3*) MOGHISSI, A. A. and C. R. PORTER. Tritium in surface waters in the United States, 1966. Radiol Health Data Rep 9:337–339 (July 1968).

(*4*) INTERNATIONAL COMMISSION OF RADIOLOGICAL UNITS AND MEASUREMENTS. Recommendations of the International Commission of Radiological Units and Measurements, National Bureau of Standards Handbook 85 (1964).

The Determination of Accuracy and Precision Limits for Tritium in Water

Forrest E. Knowles, Jr. and *Edmond J. Baratta*

Tritium is a radionuclide that is becoming of increasing concern to the general population. Since the advent of nuclear testing and construction and operation of nuclear power plants and nuclear fuel reprocessing plants, tritium has been introduced into the environment in greater quantities. As a result many Federal, State, local and private agencies have begun analyses of this radionuclide on a routine basis, while others are preparing for this capability. Reactor effluents, municipal water supplies, and rivers near nuclear power reactors and reprocessing plants are among the major environmental media for examination. A procedure of quality control to assist the laboratories in producing accurate and comparable data is necessary. Also this quality control service can help assess the equipment and sample preparation capabilities of the various laboratories.

The Analytical Quality Control Service (AQCS) has attempted to assist in this endeavor by establishing a standard deviation (σ) for tritium analyses and setting up control limits. From these criteria, a laboratory can judge its performance in various experiments and take appropriate corrective action if needed.

This paper describes the procedure for determining a standard deviation for tritium and the results of a technical experiment for validating σ.

Experimental procedure

For the past 2 years, the AQCS has been conducting a tritium cross-check water program with the three EPA laboratories and two other agencies interested in this analysis. The technical experiment consisted of triplicate analyses by each of the participating laboratories on three water samples containing known concentrations (not disclosed to the laboratories) of tritium.

Water samples (A, B, and C) were prepared by adding calibrated solutions of tritium to three 1-liter portions of distilled water. Cobalt-60 was added to sample B as a low energy beta interference. The activity of each sample solution is shown in table 1.

Table 1. Radioactivity of the sample solutions

Sample	Tritium[a] (nCi/liter)	Cobalt-60 (nCi/liter)
A	51.8	—
B	146.4	15.0
C	331.4	—

[a] The accuracy of the standards was within ± 2 percent.

The laboratories were sent 50 ml of each sample. They were instructed to perform triplicate analyses using their routine procedure and to forward these results to the AQCS. The AQCS determined the average of each laboratory's analyses and returned a listing of these averages to the laboratories. An example of this report for a particular month is shown in table 2.

Table 2. Tritium analysis—mean values for three determinations

Laboratory	Tritium concentration (nCi/liter)
A	160.7
B	170.0
C	173.7
D	183.0
E	170.0
Known	177.8
Range	22.3

Discussion

After a suitable amount of data was accumulated, a standard deviation value of 10 percent for a single determination was experimentally set by comparing each laboratory's values to the actual concentration in the sample and studying laboratory precision over an extended period of time (1½ years). This σ applied to all tritium values greater than 5 nCi/liter. Using a standard deviation value of 10 percent for a single determination, the control limits were calculated for the mean value of replicate tritium determinations as follows:

$$\pm WL \text{ (Warning Level)} = 2\sigma_m$$
$$\pm CL \text{ (Control Limit)} = 3\sigma_m$$

$$\sigma_m = \frac{\sigma}{\sqrt{N}} \qquad \ldots(1)$$

where:

σ_m = the effective standard deviation for the mean.

σ = the established standard deviation for a single determination (10 percent of the known), and

N = the number of replicate determinations.

These control limits were then applied to the AQCS cross-check program. A typical report is shown in table 3. Mean values deviating from the known value by an amount less than the warning level were considered acceptable. Those between warning level and control limit were conditionally acceptable. Mean values deviating from the known value by an amount greater than the control limit were considered unacceptable, and an investigation should be made by the laboratories involved.

Table 3. Tritium analysis—control limits for three determinations

Laboratory	Tritium concentration mean value (nCi/liter)
A	47.3
B	46.4
C	47.0
D	47.1
E	54.0
F	53.3
Known	50.4
± WL	5.8
± CL	8.7

In order to test the standard deviation for the mean value, a technical experiment was conducted in cooperation with 16 participants including the EPA laboratories. Results of the experiment were compared to known values by the use of control charts as shown in figure 1.

The mean values were used for the data analyses because the range values did not vary significantly between laboratories and were well within the warning levels. All laboratory mean values were included. The data are presented in figures 2 to 4.

Analysis of data for precision was accomplished by comparing the standard deviation of each laboratory's individual results for each sample with the 10 percent standard deviation values established by the AQCS as shown in table 4.

Table 4. Data analysis of technical experiment-precision

Sample	Known σ and (percent)	Laboratory σ[a] and (percent)
A	5.18 (10)	7.7 (14)
B	14.6 (10)	25.1 (16)
C	33.1 (10)	56.0 (16)

[a]

$$\sigma = \sqrt{\frac{\Sigma(x_i - \bar{x})^2}{N - 1}}$$

where: N = number of laboratories,
$\bar{x}$ = average of sample activity of all analyses.
x_i = individual laboratory results.

MEAN CHART	Sample A– ^3H = 51.8 nCi/liter, N = 3	
$\bar{\bar{X}} + \frac{3\sigma}{\sqrt{N}}$	60.8	Upper Control Limit (UCL)
$\bar{\bar{X}} + \frac{2\sigma}{\sqrt{N}}$	57.8	Upper Warning Level (UWL)
$\bar{\bar{X}}$	51.8	Standard Mean (Expected Value)
$\bar{\bar{X}} - \frac{2\sigma}{\sqrt{N}}$	45.8	Lower Warning Level (LWL)
$\bar{\bar{X}} - \frac{3\sigma}{\sqrt{N}}$	42.8	Lower Control Limit (LCL)

RANGE CHART	Sample A– ^3H = 51.8 nCi/liter, N = 3	
$D_4 \bar{R}$	22.6	Upper Control Limit (UCL)
$\bar{R} + \frac{2}{3}(D_4 \bar{R} - \bar{R})$	18.0	Upper Warning Level (UWL)
$\bar{R} = d_2\sigma$	8.8	Mean Range (Expected Range)
$D_3 \bar{R}$	0	Lower Limit (LL)

Figure 1. Examples of "Mean and Range Control Charts"-
d_2, D_3, and D_4 are values taken from standard statistical tables prepared for the construction of range control charts and based on the value of N

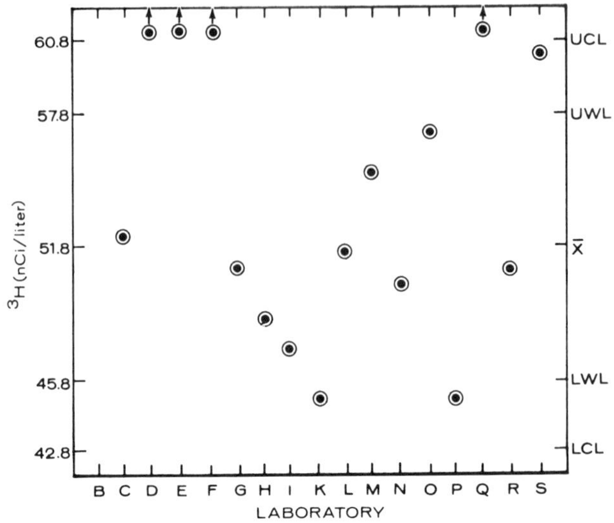

Figure 2. "Mean Chart," sample A, 51.8 nCi/liter, N = 3

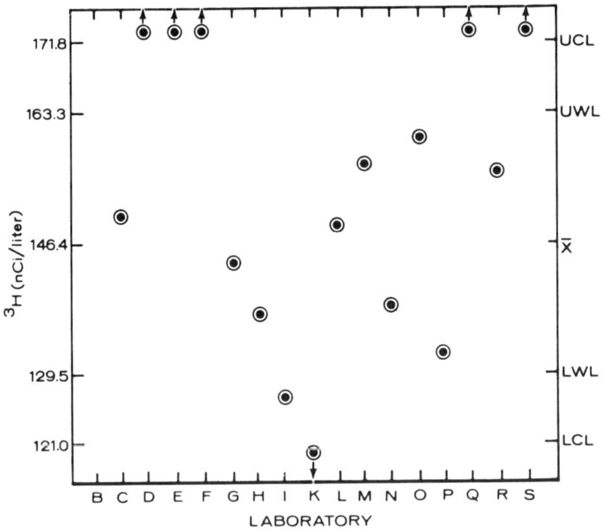

Figure 3. "Mean Chart," sample B, 146.4 nCi/liter, N = 3

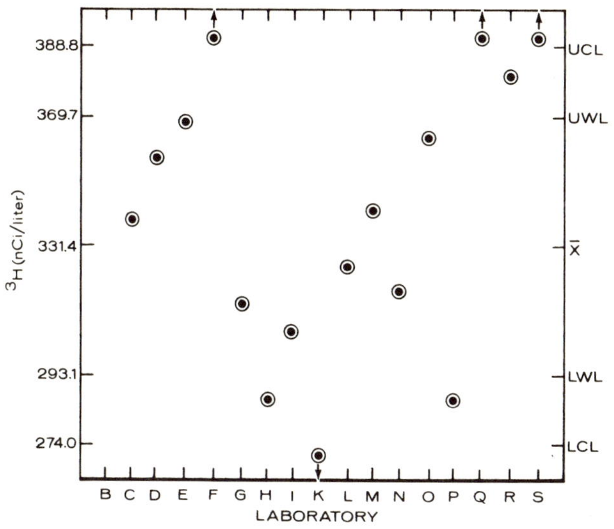

Figure 4. "Mean Chart," sample C, 331.4 nCi/liter, N = 3

To determine the accuracy of the tritium procedures, the difference between the mean values and the known values was compared to the established standard deviation for a single determination and for three determinations. The results are shown in tables 5 and 6.

Table 5. Data analysis of technical experiment

Sample	Amount present (nCi/liter)	Average of means (nCi/liter)
A	51.8	54.3
B	146.4	154.7
C	331.4	342.5

Table 6. Data analysis of technical experiment—accuracy

Sample	Difference (nCi/liter)	Percent[a]	σ[b] (nCi/liter)	σ_m[c] (nCi/liter)
A	2.5	4.8	5.18	3.0
B	8.3	5.7	14.6	8.4
C	11.1	3.3	33.1	19.1

[a] Percent $= \frac{\text{difference}}{\text{average amount present}}$

[b] σ = standard deviation for tritium, single determination (10 percent)

[c] σ_m = standard deviation for tritium, three determinations $\left(\frac{\sigma}{\sqrt{N}}\right)$

Conclusions

The calculated σ for the three samples is larger than the one established for tritium in the precision analysis (table 4) because a few participants had values differing from the known greatly in excess of the control limits. The data were recalculated excluding three laboratories, and the value came close to 10 percent. (A = 10.5, B = 11.2, C = 11.4). The relative σ does not seem to be dependent upon sample activity.

The analysis of accuracy (table 6) shows that the percent difference between the average amount present and average amount found is comparable to the established standard deviation. In some cases, these values are much better than expected.

From the data shown in the tables and figures, it appears that the established 10 percent standard deviation is valid. Values lower than 50 nCi/liter (25–30 nCi/liter) have been used in the cross-check program with similar results. Levels nearer 1–5 nCi/liter will be investigated in future studies, as some laboratories in the program routinely count samples at these lower levels. Due to the greater influence of background at 5 nCi/liter or less, the 10 percent standard deviation will have to be revised upward.

Gross Radioactivity in Surface Waters of the United States, January 1971

Office of Water Programs
Environmental Protection Agency

The monitoring of gross radioactivity in surface waters of the United States was begun in 1957 as part of the Water Pollution Surveillance System (formerly National Water Quality Network) of the U.S. Public Health Service. Currently, the program is operated by the Environmental Protection Agency, Office of Water Programs. Regional Offices of the Environmental Protection Agency are responsible for the collection of samples and entering of the analytical findings into STORET, the Office of Water Programs' computerized data storage and retrieval system. Radioactivity analyses are performed in the centralized radioactivity laboratories of the Office of Water Programs (Cincinnati, Ohio).

The regular reporting of gross radioactivity data in *Radiological Health Data and Reports* was terminated with the publication of data for the month of October 1968 (April 1969 issue). With this publication of data for January 1971, it is intended to reinitiate this activity. The unpublished data for the time interval of November 1968 through December 1970 will be the the subject of a future summary article.

Table 1 presents the gross alpha and beta results for samples collected from 12 rivers during January 1971. The analytical procedures used for determining gross alpha and beta radioactivity are described in the 13th Edition of "Standard Methods for the Examination of Water and Wastewater" (*1*). Results are reported for the date of counting and are not corrected by extrapolation to the date of collection. The sensitivity in counting is that defined by the National Bureau of Standards, Handbook 86 (*2*) and is calculated to be <0.2 pCi/liter for gross alpha and <1 pCi/liter for gross beta measurements.

Table 1. Gross radioactivity in U.S. surface waters, January 1971

River and station	Number of grab samples[a]	Gross alpha radioactivity (pCi/liter)		Gross beta radioactivity (pCi/liter)	
		Suspended solids	Dissolved solids	Suspended solids	Dissolved solids
Clinch River:					
Kingston, Tenn.	4	<0.2 (< .2, <0.2)	<0.2 (< .2, < .2)	<1 (<1, 1)	10 (7, 14)
Great Miami:					
Elizabethtown, Ohio	1	1	2	1	11
American Material Bridge, Ohio	1	< .2	2	3	8
Fairfield, Ohio	1	1	2	1	11
Sellars Road Bridge, Ohio	1	< .2	2	1	7
Kentucky River:					
Lock and Dam No. 1	1	2	< .2	11	3
Licking River:					
Covington, Ky.	1	2	< .2	8	3
Little Miami River:					
Cincinnati, Ohio	1	< .2	2	2	10
Mississippi River:					
Burlington, Iowa	1	< .2	1	1	7
Ohio River:					
Markland Dam	1	< .2	2	5	5
Cincinnati, Ohio	4	<1 (< .2, 1)	< .2	6 (2, 11)	4 (3, 5)
Portsmouth, Ohio	1	3	2	10	4
Ironton, Ohio	1	< .2	1	2	3
Roanoke River:					
John Kerr Dam, Va.	3	< .2 (< .2, < .2)	< .2 (< .2, < .2)	<1 (<1, 1)	4 (4, 4)
St. Lawrence River:					
Massena, N.Y.	3	< .2 (< .2, < .2)	< .2 (< .2, 1)	<1 (<1, 1)	6 (6, 6)
St. Mary River:					
Sault Ste. Marie, Mich.	1	< .2	< .2	<1	2
Susquehanna River:					
Holtwood, Pa.	1	< .2	< .2	2	3
Whitewater River:					
State Line (Ohio-Ind.)	1	< .2	4	<1	5

[a] Where more than one sample is analyzed, the minimum and maximum values are in parentheses.

REFERENCES

(*1*) AMERICAN PUBLIC HEALTH ASSOCIATION; AMERICAN WATER WORKS ASSOCIATION AND WATER POLLUTION CONTROL FEDERATION. Standard methods for the examination of water and wastewater, 13th Edition, New York, N.Y. (1971).

(*2*) U.S. DEPARTMENT OF COMMERCE. Radioactivity, recommendations of the International Commission on Radiological Units and Measurements (1962), NBS Handbook 86 (November 29, 1963).

Gross Radioactivity in Surface Waters of the United States, February 1971

Office of Water Programs
Environmental Protection Agency

The monitoring of gross radioactivity in surface waters of the United States was begun in 1957 as part of the Water Pollution Surveillance System formerly National Water Quality Network) of the Public Health Service. Currently, the program is operated by the Environmental Protection Agency, Office of Water Programs. Regional offices of the Environmental Protection Agency are responsible for the collection of samples and the entering of the resulting data into the analytical storage and retrieval system. Radioactivity analyses are performed in the centralized radioactivity laboratories of the Office of Water Programs (Cincinnati, Ohio).

Table 1. Gross radioactivity in U.S. surface waters, February 1971

River and station	Number of grab samples[a]	Gross alpha radioactivity (pCi/liter)		Gross beta radioactivity (pCi/liter)	
		Suspended solids	Dissolved solids	Suspended solids	Dissolved solids
Clinch River: Kingston, Tenn.	3	(<0.2, <0.2) <0.2	(<0.2, 2) <0.8	(<1, 1) <1	(9, 11) 10
Ohio River: Cincinnati, Ohio	1	<0.2	<0.2	1	3
Roanoke River: John Kerr Dam	2	(<0.2, <0.2) <0.2	(<0.2, 1) <0.6	(1, 1) <1	(4, 4) 4
St. Lawrence River: Massena, N.Y.	3	(<0.2, <0.2) <0.2	(<0.2, <0.2) <0.2	(<1, <1) <1	(6, 6) 6

[a] Where more than one sample is analyzed, the minimum and maximum values are in parentheses.

Table 1 presents the gross alpha and beta results for samples collected from four rivers during February 1971. The analytical procedures used for determining gross alpha and beta radioactivity are described in the 13th Edition of "Standard Methods for the Examination of Water and Wastewater" (*1*). Results are reported for the date of counting and are not corrected by extrapolation to the date of collection. The sensitivity in counting is that defined by the National Bureau of Standards, Handbook 86 (*2*) and is calculated to be <0.2 pCi/liter for gross alpha and <1 pCi/liter for gross beta radioactivity measurements.

REFERENCES

(*1*) AMERICAN PUBLIC HEALTH ASSOCIATION; AMERICAN WATER WORKS ASSOCIATION AND WATER POLLUTION CONTROL FEDERATION. Standard methods for the examination of water and wastewater, 13th Edition. New York, N.Y. (1971).

(*2*) U.S. DEPARTMENT OF COMMERCE. Radioactivity, Recommendations of the International Commission on Radiological Units and Measurements (1962), NBS Handbook 86 (November 29, 1963).

EVALUATION OF ION-EXCHANGE SURVEILLANCE SAMPLER FOR ANALYZING RADIOACTIVE LIQUID EFFLUENTS

H. L. KRIEGER and G. W. FRISHKORN

INTRODUCTION

To be able to satisfy the increasing power demands of industry and the consumer, the production of electrical energy may have to come from nuclear sources to a much greater extent. In terms of electricity generation, nuclear power may exceed 25% of total power by 1980.[1] This predicted expansion will require more stringent nuclear safeguards plus updated information on population exposure levels and on the critical pathways of radionuclide released during reactor operations. This study, which field tested a surveillance sampling device, was designed to identify and measure radionuclides in discharges, and to provide some of this needed information.

The nuclides produced in water-cooled reactors have been enumerated in a comprehensive survey.[2] Reactor station progress reports and publications in the scientific literature have also identified most of the nuclides in nuclear reactor aqueous discharges, but little data are available as to the concentrations of specific nuclides in discharges. These activity levels, which are at present in the 10^{-1} pCi/l. level at the point of release, could increase significantly as reactor operations expand. Therefore, the need exists today to measure with confidence the activity of these nuclides in reactor discharges to determine whether pollution controls should be instituted to minimize a possible population radiation exposure problem.

Radiochemical analyses for nuclides in the pCi/ml range are normally performed on sample aliquots of 500 ml or less. However, the lower levels encountered in reactor effluents require analysis of much larger samples. Ion-exchange resins have been very effective in analytical chemistry separations,[3] and have been used extensively in water purification, radionuclide concentration, and specific ion removal.

River waters which are normally used for cooling, contain between 10 and 250 mg/l. $CaCO_3$ (ppm hardness) with the average around 150–200.[4] This calcium concentration, which is significantly greater than that of any other cation in water, would be the limiting factor in the design of the surveillance sampler. Since at least 50 l. of effluent should be concentrated to assure satisfactory levels for analysis, the surveillance sampler had to be designed to collect this minimum volume of a 200 ppm hardness water. This would require sufficient ion-exchange to collect all the radioactive cations and anions present, and to prevent calcium breakthrough.

The effectiveness of this surveillance sampler has been evaluated as a rapid and reliable method for concentrating and measuring the activities in aqueous reactor discharges.

LABORATORY STUDIES

The surveillance sampler consisted of strongly acidic cation resin in the H^+ ionic form (Dowex 50 X-8)* and strongly basic anion resin in the Cl^- form (Dowex 1 X-8). The total exchange capacity rating of the wet cation resin (1.8 mequiv./ml) was substantiated by a laboratory study in which 25 l. of 340 ppm hardness water was passed through three sections of wet cation resin, 35 g each. Since >95% of the 8500 mg Ca^{++} as $CaCO_3$ was exchanged on the top two sections, it indicated that retention would be >8100 mg/70 g or ~116 mg $CaCO_3$/g resin. A column containing 1000 g cation resin would retain at least 75,000 mg $CaCO_3$ in the top two-thirds of the column. A combined cation and anion resin column of this size would therefore concentrate and completely retain the cation and anion activity present in 300 l. of a water with a hardness of 250 ppm.

The nuclide retention efficiency of the resins in the effluent sampler is shown in Table 1. The tracers, in 100 l. of a 200 ppm hardness water, were passed through the three sections of the cation resin only. The water containing the ^{131}I was passed through the two sections of anion resin only. The results disclose that at least 95% of each nuclide was retained in the top section of the column.

The resin arrangement of the sampler was determined with tracer solutions of ^{85}Sr, ^{137}Cs, ^{60}Co, ^{140}Ba-^{140}La and ^{131}I in waters of hardnesses varying from 150 to 250 ppm $CaCO_3$. These solutions, containing at least 10,000 dpm of each tracer, were passed through the surveillance sampler at a flow rate of 100 ml/min. This resin column, 400 × 80 mm i.d., was filled with three sections of moist cation resin (350 g each), and two

* Mention of commercial products does not constitute endorsement by the Division of Environmental Radiation.

Table 1. *Retention efficiency of the effluent sampler for tracers in* 100 l. *of* 200 ppm *hardness water*

	% Retained			
	^{60}Co	^{85}Sr	^{137}Cs	^{131}I
Cation section				
1	95.0	98.5	96.5	—
2	4.0	1.5	3.5	—
3	1.0	—	—	—
Anion section				
1	—	—	—	96.5
2	—	—	—	3.5

Resin column dimensions: 200 × 30 mm i.d. Cation resin: x-8 cross linkage; 50–100 mesh. Anion resin: x-8 cross linkage; 50–100 mesh. Flow rate: 40–50 ml/min.

sections of moist anion resin (400 g each). Perforated plastic spacers between each section permitted separation of the resin for subsequent elution without restricting flow through the column.

About 10–20% of the ^{131}I was physically attached to the cation resins when all three cation sections preceded the anion sections in the sampler. The arrangement which was finally adopted involved a top cation section followed by an anion section, then two cation sections and a final anion section. In this manner, nearly all the activity was concentrated in the first two resin sections, and minimized anion loss resulting from long contact with cation resin. A schematic drawing of the apparatus is shown in Fig. 1.

The elution of the radionuclides and calcium from the resin was studied using the smaller column containing three-35 g sections of cation resin. After 40 l. of a 150 ppm hardness water had been put through the column, each resin section was eluted with 6 N HCl until no calcium was evident in the eluate. The results indicated that 150 ml of 6 N HCl per 35 g cation resin would quantitatively elute the calcium and the exchanged radionuclides.

FIELD INVESTIGATIONS

At the reactor site, the effluent surveillance samplers were set up in the coolant-canal;

one sampler in the inlet to the plant and the other in the plant outlet at the end of the discharge canal. When reactor operations were discharging high-conductivity waste, the mixed outlet was pumped into a 55-gal drum, and the inlet water was collected in another drum.

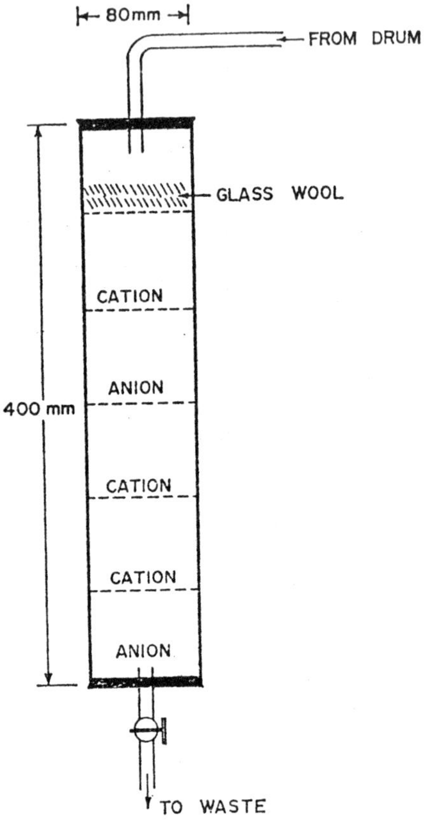

FIG. 1. Surveillance sampler (schematic).

The intake and discharge canal waters were siphoned into respective samplers at a flow rate of 100 ml/min. When ~150 l. had been concentrated, the five resin sections in each column were separated and analyzed spectrometrically with a 10- × 10-cm cylindrical NaI(Tl) gamma-ray detector. Spectra of the top cation and top anion sections are shown in Figs. 2 and 3. Photopeaks evident in these gamma scans are ^{58}Co, ^{60}Co, ^{134}Cs and ^{140}Ba-^{140}La in the cation resin and ^{131}I in the anion resin. The resins were measured again after an interval of one week to confirm gamma spectral identification by the decay of the gamma-ray peaks for ^{140}Ba-^{140}La and ^{131}I in particular. Gamma-ray spectra of the other resin sections very closely resembled instrument background, giving additional evidence that activity was quantitatively retained on the top resin sections.

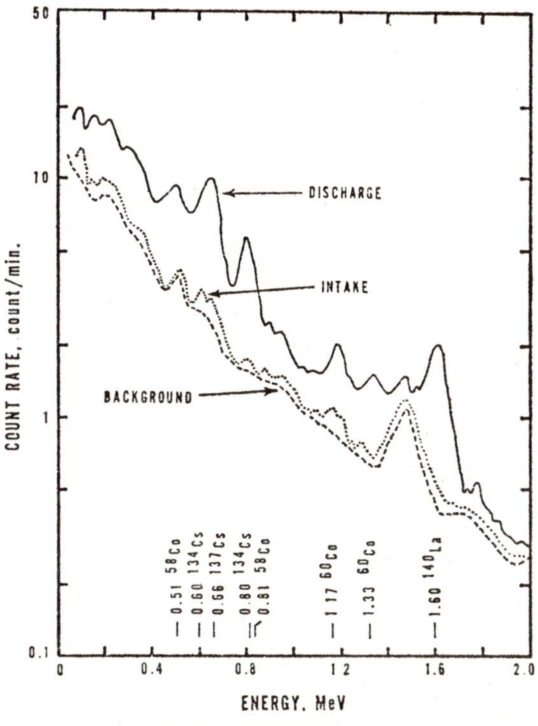

Fig. 2. Gamma-ray spectra of cation-exchange resins in intake and discharge coolant-water canals.

Each cation-exchange resin was eluted with 1500 ml 6 N HCl at a flow rate of 50 ml/min. Aliquots were analyzed in duplicate for ^{134}Cs, ^{137}Cs, ^{89}Sr, ^{90}Sr, ^{58}Co, ^{60}Co and ^{140}Ba-^{140}La by radiochemical separation. The aliquot size was usually 200–500 ml, depending upon the amount of calcium in the sample.

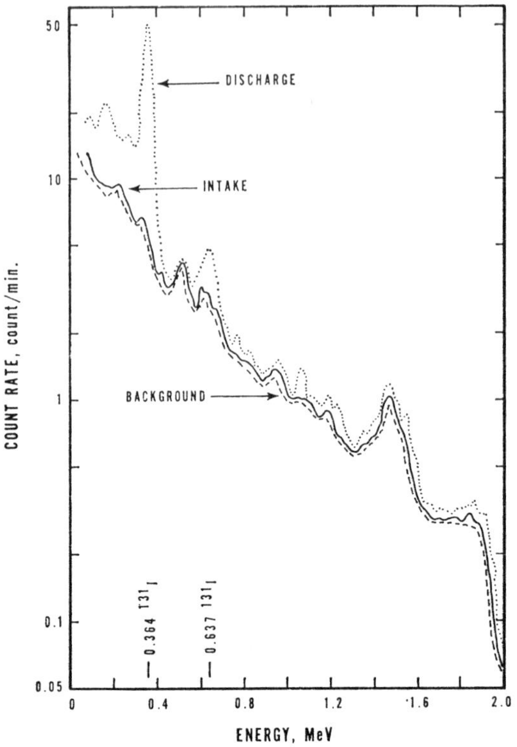

FIG. 3. Gamma-ray spectra of anion-exchange resins in intake and discharge coolant-water canals.

Since 1500 ml of acid eluate was equivalent to about 150 l. of reactor coolant canal water, these aliquots represented a 20–50 l. sample and enabled satisfactory interpretation of results. The values obtained during one surveillance field trip to a boiling water reactor nuclear power station are listed in Table 2. The activities determined in the inlet can be attributed to fallout while the difference between outlet and inlet values are considered station effluent. The ^{131}I activities in the two anion resin sections were calculated from the respective spectra on the basis of gamma efficiencies obtained for this geometry. Decay measurements corroborated the results obtained.

These reactor effluent activities can be compared to the predicted values which were obtained by analyzing the liquid waste before dilution.[5] The concentrations to be expected in the discharge canal are found to depend not only on flow and dilution, but also on uptake by flora, fauna and sediment. If the activity in intake and effluent samples are determined with confidence and accuracy using the surveillance column, the results become more meaningful because they are based on analysis of larger sample volumes.

The surveillance sampler has been field tested at one reactor site several times, and each time modifications have been made to improve either the collection or analysis phase. The results of these surveillance studies on reactor effluents, which can be used to assess population radiation exposure, have shown that the concentration of critical nuclides in the reactor effluent tested has been below the AEC limits designated for unrestricted areas.[6]

Table 2. Activity levels in reactor inlet and outlet as determined with ion-exchange surveillance sampler

	pCi/l.	
	Inlet	Outlet
^{89}Sr	<0.5	<0.5
^{90}Sr	0.47	0.81
^{58}Co	<0.1	<0.1
^{60}Co	0.2	0.2
^{134}Cs	0.1	0.2
^{137}Cs	0.1	0.2
^{140}Ba-^{140}La	<0.1	0.2
^{131}I*	<0.02	0.8

* The ^{131}I values were determined from the gamma-ray spectrum; all others were the result of radiochemical analysis of leach aliquots. Resin sampler dimensions: 400 × 80 mm i.d. Cation resin: x-8 cross linkage; 50–100 mesh. Anion resin: x-8 cross linkage; 50–100 mesh. Flow rate: 100 ml/min.

SUMMARY

An ion-exchange surveillance sampler has been used in field operations to analyze for the nuclides ^{58}Co, ^{60}Co, ^{80}Sr, ^{90}Sr, ^{134}Cs, ^{137}Cs, ^{131}I and ^{140}Ba-^{140}La at 10^{-1} pCi/l. levels in reactor effluent water at point of release. The use of two such units at both the inlet to and outlet from the plant makes possible the determination of the activity levels resulting from reactor operation. Although the concentrations of separated radionuclides were measured by conventional radiochemical procedures, the results at the ion-exchange techniques were at least an order of magnitude more sensitive because of the larger sample volume employed.

The design and resin arrangement of this unit resulted from laboratory studies using specified nuclides and water characteristics. Should the nature or composition of future reactor releases warrant a change in these specifications, the surveillance sampler could be modified to conform to these needs.

Acknowledgement—The assistance of ELEANOR MARTIN, who handled the analytical and counting data, is gratefully acknowledged.

REFERENCES

1. *Chem. Engng News* **48,** 16 (1970).
2. J. E. MOUNTAIN, L. E. ECKART and J. H. LEONARD, Survey of individual radionuclide production in water-cooled reactors. Summary Rep., Contract PH 86-67-218, Nucl. Sci. Engng Dept., Univ. Cincinnati, May (1968).
3. O. SAMUELSON, *Ion Exchangers in Analytical Chemistry*, John Wiley, New York (1953).
4. Public water supplies of 100 largest cities in U.S., 1962. U.S. Geological Survey Water Supply Paper 1812, Washington, D.C. (1962).
5. B. KAHN *et al.*, Radiological surveillance studies at a boiling water nuclear power reactor. PHS Rep., BRH/DER 70-1, March (1970).
6. USAEC, Standards for protection against radiation. Title 10, Code Fedl Regulat., Part 20 (U.S. Govt Printing Office, Washington, D.C. (1965)).

Statistical Limits of Accuracy and Precision of Gross Alpha and Beta Radioactivity Measurements in Water

Edmond J. Baratta and *Forrest E. Knowles, Jr.*

A technical experiment designed to determine the detection limits, accuracy and precision of gross alpha and beta radioactivity measurements in water was conducted by the Environmental Protection Agency's Analytical Quality Control Service at the Northeastern Radiological Health Laboratory. These analyses are normally used as a "screening" technique. Since they are used as such, it is desirable to know the limits of accuracy and precision. These measurements are routinely made by the State Health Departments and Power Companies around areas proposed as reactor sites and fuel reprocessing plants in order to establish a background survey. When the nuclear facilities become operational, the gross alpha and beta measurements are made to monitor the effluent from the plants. These are performed routinely since they are fairly inexpensive to make, as compared with specific analyses of the radionuclides present.

The Public Health Service also has found these measurements very useful as a screening technique (*1*). For guidance on drinking water, it has established an alpha limit of 3 pCi/liter and a beta limit of 10 pCi/liter before specific analyses are made. If gross alpha radioactivity is above 3 pCi/liter, then a specific analysis is made for radium-226. If gross beta radioactivity is above 10 pCi/liter, then a specific analysis is made for strontium-90.

Preparation of samples and description of test

Test materials for this experiment consisted of a standard solution of alpha radioactivity, a standard solution of beta radioactivity and tap water for a source of dissolved solids. The alpha activity was natural uranium and the beta activity was cesium-137. These solutions were dilutions of stock activity prepared in acid solution so as to eliminate loss of activity on the walls of vials. The supplied tap water was drinking water from Woburn, Mass., with a solids content of 350 mg/liter. Two pairs of samples containing different levels of radioactivity were prepared from the alpha and beta solutions. The concentrations in these samples are described in table 1.

Five ml portions of each alpha and beta solution and 4 liters of the drinking water were forwarded to the participants. In addition, cesium-137 and natural uranium calibration solutions were forwarded for use in preparation of absorption curves. The participants were instructed to pipet 1 ml of each alpha solution and 1 ml of each beta solution into separate appropriate aliquots of supplied water. A single analysis was to be performed for each sample.

Table 1. Radioactivity in test samples[a]

Radioactivity	Pair 1		Pair 2	
	Sample R	Sample S	Sample T	Sample U
Alpha (pCi natural U/sample[b])	37.9	25.4	100.2	80.5
Beta (pCi ^{137}Cs/sample)	62.5	47.7	130.0	156.0

[a] The accuracy of the standards were within ±2 percent.
[b] pCi/sample refers to picocuries per 1 milliliter of supplied alpha and beta activity.

Methodology

Procedures used to measure the gross alpha and gross beta activities were taken from the 12th Edition of the American Public Health Association (APHA) Standard Methods (*2*). Analysis of gross alpha and beta radioactivity is as follows: The specific conductance is measured first to help select the proper sample volume. The sample is prepared by (1) adding sample directly to a tared counting pan with evaporation just below the boiling temperature, or (2) placing the sample in a pyrex beaker, adjusting the pH to 4–6, and evaporating to near dryness. The residue is transferred to a tared counting pan using water and dilute acid washings.

The counting pan is dried in an oven at 103–105° C, cooled in a desiccator, weighed, and kept dry until counted. The sample is counted in an internal proportional counter or other appropriate counting instrument.

Analysis of data

The results of the gross alpha test are presented in table 2. The results of the gross beta test are presented in table 3. The design and evaluation of these tests are based in part on Youden's guide on "Statistical Techniques for Collaborative Tests" (3). This design assumes that the internal precision is not different at each laboratory, and the differences which usually exist between laboratories are associated with laboratory systematic errors.

Table 2. Gross alpha radioactivity in water

Laboratory code[a]	Results (pCi/sample)			
	Sample R	Sample S	Sample T	Sample U
E	40.4	28.3	97.3	78.8
G	14.9	10.6	34.9	31.7
I	34.0	23.8	91.4	73.8
J	35.5	21.5	90.7	77.5
M	35.3	26.4	83.2	65.2
N	57.8	34.6	155.7	112.0
O	24.2	20.5	44.7	25.0
Q	34.8	25.3	85.4	76.7
R	33.0	22.8	65.0	67.6
S	15.1	12.3	53.4	28.4
T	36.1	24.0	97.3	78.8
V	48.3	29.5	135.7	107.8
X	42.1	27.2	120.8	89.8
Y[b]	4.63	3.58	15.6	11.3
Average	34.7	23.6	88.9	70.2
Range	42.9	24.0	120.8	87.0
Known	37.9	25.4	100.2	80.5

[a] Laboratory code P and AA did not participate in this analysis.
[b] Not included in range or average.

There are three distributions which are of interest: S_d the combined effect of the distribution of systematic errors and the random errors associated with the individual results. S_a a measure of how successful repeat determinations can be made (degree of random error). S_b—a measure of the displacement (or change in bias) that may be encountered as one changes laboratories (or degree of laboratory systematic error). The distribution has the following relationship:

$$S_d^2 = S_b^2 + S_a^2 \qquad \ldots(1)$$

The actual data yield S_d^2 and S_a^2. S_b^2 is determined by difference. The following formulas were used to make the various calculations:

$$S_d = \sqrt{\frac{(T_i - \bar{T})^2}{2(n-1)}} \qquad \ldots (2)$$

T_i = Total of the paired samples for the "i"th lab

$\bar{T}$ = average total

n = number of totals

$$S_a = \sqrt{\frac{(D_i - \bar{D})^2}{2\,(n-1)}} \qquad \ldots (3)$$

D_i = difference between paired samples for the "i"th lab

$\bar{D}$ = average difference

n = number of differences

An estimate of the method accuracy is obtained by the difference between the average known concentration and the average found concentration for each pair. The relative accuracy is calculated by dividing this difference by the average known concentration. Relative precision for the method is estimated by dividing the random error (S_n) by the average amount found for each pair. This statistical analysis for the gross alpha and beta radioactivity is presented in tables 4 and 5, respectively. The summary for gross alpha and beta radioactivity is presented in table 6.

Discussion of results

Accuracy and precision for this method were determined using the data from all contributing laboratories except in the case of gross alpha radioactivity where one outlying laboratory was not included. For both alpha and beta determinations,

Table 3. Gross beta radioactivity in water

Laboratory code	Results (pCi/sample)			
	Sample R	Sample S	Sample T	Sample U
E	67.5	54.1	146.0	168.8
G	39.2	46.6	54.5	116.0
I	43.2	39.6	108.5	139.6
J	62.0	49.6	125.5	155.3
M	93.5	78.4	217.0	234.0
N	78.8	65.2	170.7	200.4
O	50.7	34.4	111.2	119.7
P	68.8	45.5	131.0	156.0
Q	66.8	51.7	135.3	157.1
R	61.6	52.6	125.4	151.2
S	64.2	49.5	126.3	142.8
T	65.4	51.5	127.4	155.0
V	55.4	45.4	118.2	152.7
X	59.9	48.6	128.2	154.4
Y	49.3	39.0	103.0	122.0
AA	59.5	47.4	134.9	152.0
Average	61.6	49.9	128.9	154.8
Range	54.3	44.0	162.5	118.0
Known	62.5	47.7	130.0	156.0

Table 4. Data analysis, gross alpha radioactivity in water

Laboratory code[a]	Difference (R−S)	Total (R+S)	Difference (T−U)	Total (T+U)
Known	12.5	63.3	19.7	180.7
E	12.1	68.7	18.5	176.1
G	4.3	25.5	3.2	66.6
I	10.2	57.8	17.6	165.2
J	14.0	57.0	13.2	168.2
M	8.9	61.7	18.0	148.4
N	23.2	92.4	43.7	267.7
O	3.7	44.7	19.7	69.7
Q	9.5	60.1	8.7	162.1
R	10.2	55.8	−2.6	132.6
S	2.8	27.4	25.0	81.8
T	12.1	60.1	18.5	176.1
V	18.8	77.8	27.9	243.5
X	14.9	69.3	31.0	210.6
Total	144.7	758.3	242.4	2,068.6
Average	11.1	58.3	18.6	159.1
S_a^2	17.1		72.0	
S_a	4.1		8.5	
S_d^2		167.3		1,901
S_d		12.9		43.6

[a] Laboratory Y was not included in data analysis.

there were some laboratories which could have been excluded from the data analysis due to their deviation from the known but were included to keep the number of laboratories at a maximum.

Table 5. Data analysis, gross beta radioactivity in water

Laboratory code	Difference (R − S)	Total (R + S)	Difference (U − T)	Total (U + T)
Known	14.8	110.2	26.0	286.0
E	13.4	121.6	22.8	314.8
G	−7.4	85.8	61.5	170.5
I	3.6	82.8	31.1	248.1
J	12.4	111.6	29.8	280.8
M	15.1	171.9	17.0	451.0
N	13.6	144.0	29.7	371.1
O	16.3	85.1	8.5	230.9
P	23.3	114.3	25.0	287.0
Q	15.1	118.5	21.8	292.4
R	9.0	114.2	25.8	276.6
S	14.7	113.7	16.5	269.1
T	13.9	116.9	27.6	282.4
V	10.0	100.8	34.5	270.9
X	11.3	108.5	26.2	282.6
Y	10.3	88.3	19.0	225.0
AA	12.1	106.9	17.1	286.9
Total	186.7	1,784.9	413.9	4,540.1
Average	11.7	111.6	25.8	283.8
S_a^2	21.4		67.5	
S_a	4.6		8.2	
S_d^2		259.0		1,914.9
S_d		16.1		43.8

The gross alpha determination appears to be biased low, as observed from the differences the average amount present and the average amount found. The accuracy ranges from minus 8 percent to 12 percent and the precision ranges from 11 percent to 14 percent. In most cases the laboratories did quite well in determining the difference between the paired samples. In the absence of laboratory systematic error, S_a and S_d are estimates of the same within laboratory precision. S_a is substantially smaller than S_d for sample pairs which indicates the presence of some degree of laboratory systematic error. Possible explanations for the low bias are (1) problems in self-absorption since some laboratories experienced excessive weights of solids, and (2) incomplete drying of samples.

The gross beta results did not show any bias. The average amount present and average amount found were nearly identical for both pairs. The accuracy is about 1 percent and the precision ranges from 6 percent to 8 percent. The labora-

Table 6. Summary of results

Pair	Average amount found	Average amount present	Difference amount present-amount found	S_a	$\frac{S_a \times 100}{\text{Average amount found}}$
Gross alpha radioactivity in water:					
1	29.2	31.7	2.5	4.1	14.0
2	79.6	90.4	10.8	8.5	10.7
Gross beta radioactivity in water:					
1	55.8	55.1	−0.7	4.6	8.2
2	141.9	143.0	1.1	8.2	5.8

tories did well in determining the differences between the paired samples. Laboratory systematic errors are present in some degree since S_a is substantially less than S_d for both sample pairs. Problems in solids preparation were experienced by some of the laboratories but did not appear to be serious.

The results of the technical experiment show that this procedure for gross alpha and beta determination is acceptable as a standard method. As a screening technique, the procedure performed better than expected and could be applied to limits for specific radionuclides. The only problem appears to be in amount of solids present but if the suggested aliquot size and drying technique is followed carefully, it should not be serious.

Samples with activities around 3 pCi/liter for alpha and 10 pCi/liter for beta have not been used for a technical experiment. These levels have been used in the Analytical Quality Control Service cross-check water program. *(4)* The results of samples in that program with activities similar to the samples in this technical experiment have been comparable.

REFERENCES

(*1*) U.S. PUBLIC HEALTH SERVICE. Drinking water standards, revised 1962, PHS Publication No. 956. Superintendent of Documents, U.S. Government Printing Office, Washington, D.C. 20402 (March 1963).

(*2*) AMERICAN PUBLIC HEALTH ASSOCIATION. Water and waste water, APHA standard method, 12th Edition (1965).

(*3*) YOUDEN, W. J. Statistical techniques for collaborative tests. Association of Official Analytical Chemists, P.O. Box 540, Ben Franklin Station, Washington, D.C. 20044 (1967).

(*4*) BUREAU OF RADIOLOGICAL HEALTH. Quarterly reports for April-June 1970, July-September 1970, and October-December 1970. Northeastern Radiological Health Laboratory, Analytical Quality Central Service, Winchester, Mass. 01890.

Environmental Levels of Radioactivity at Atomic Energy Commission Installations

The U.S. Atomic Energy Commission (AEC) receives from its contractors semiannual reports on the environmental levels of radioactivity in the vicinity of major Commission installations. The reports include data from routine monitoring programs where operations are of such a nature that plant environmental surveys are required.

Releases of radioactive materials from AEC installations are governed by radiation standards set forth by AEC's Division of Operational Safety in directives published in the "AEC Manual."[1]

Summaries of the environmental radioactivity data follow for the Feed Materials Production Center and Portsmouth Area Gaseous Diffusion Plant.

[1] Title 10, Code of Federal Regulations, Part 20, "Standards for Protection Against Radiation" contains essentially the standards published in Chapter 0524 of the AEC Manual.

1. Feed Materials Production Center[2] July–December 1970

National Lead Company
Fernald, Ohio

The Feed Materials Production Center (FMPC) is operated by the National Lead Company of Ohio for the Atomic Energy Commission (AEC). The location as related to populated areas is shown in figure 1. Cincinnati and Hamilton, the larger nearby communities, are situated 20 and 10 miles from the center, respectively. Operations at this center are concerned with the processing of high-grade uranium concentrates into metallic uranium. These processes include acid digestion of the concentrates, organic phase extraction of uranyl nitrate, subsequent conversion of the uranyl nitrate to uranium oxide and tetrafluoride, reduction to uranium metal, and fabrication of the metal into fuel elements.

The FMPC also processes thorium to produce purified thorium compounds and metal. The production methods are similar to those used in producing uranium.

During the many involved reactions and processes that lead to the production of reactor fuels, liquid and airborne wastes are generated which contain varying quantities of uranium and thorium. Several in-plant methods are used to curtail their release into the environment surrounding the plant. Almost complete removal of the materials is accomplished by using dust collectors and waste treatment processses. An environmental monitoring program has been established

[2] Summarized from "Feed Materials Production Center Environmental Monitoring Semiannual Report for the Second Half of 1970" (NLCO-1079).

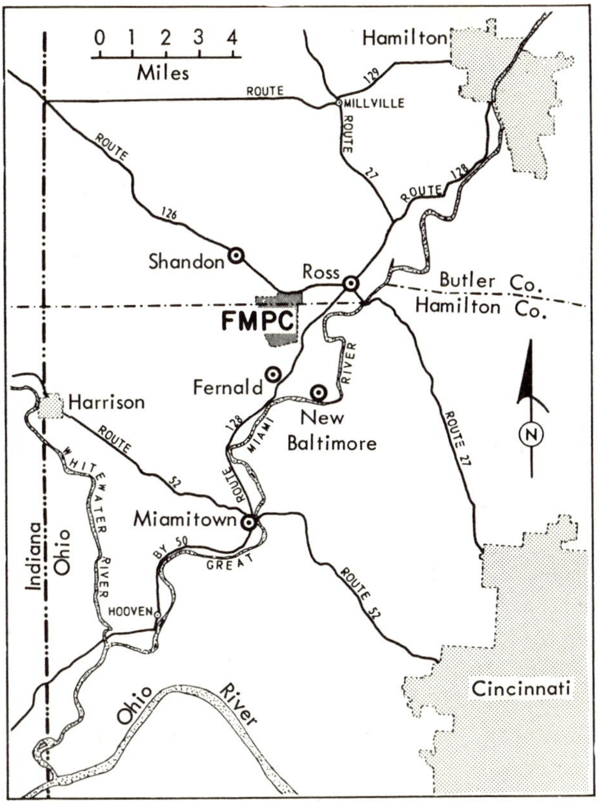

Figure 1. Area map of Feed Materials Production Center

to determine the concentrations of plant materials in the water and air outside the project.

Air monitoring

Onsite air samples are obtained from four permanent perimeter air sampling stations, located at the four corners of the production area as shown in figure 2. Samples from these perimeter stations are collected once each week and analyzed for uranium and gross alpha and gross beta radioactivity. Offsite samples are collected by a mobile air sampling unit. The location at which samples are collected is determined by local meteorological conditions on the day of sampling. Approximately 80 percent of all samples are taken downwind of the FMPC plant. Replicate samples are taken at each sampling point and averaged to obtain a representative concentration for that location. An analysis for thorium is not considered necessary because of the small amount of thorium handled in the center. Concentrations of uranium and alpha and beta radioactivity of airborne particulates sampled at onsite and offsite locations are given in table 1.

The results of sampling indicate that the concentrations at onsite locations averaged 1.6 percent, 2.1 percent, and 0.01 percent of the AEC standards for uranium, alpha radioactivity, and beta radioactivity, respectively, and the offsite concentrations averaged 1.77 percent, 1.77 percent, and 0.02 percent of the AEC standards for uranium, alpha radioactivity, and beta radioactivity.

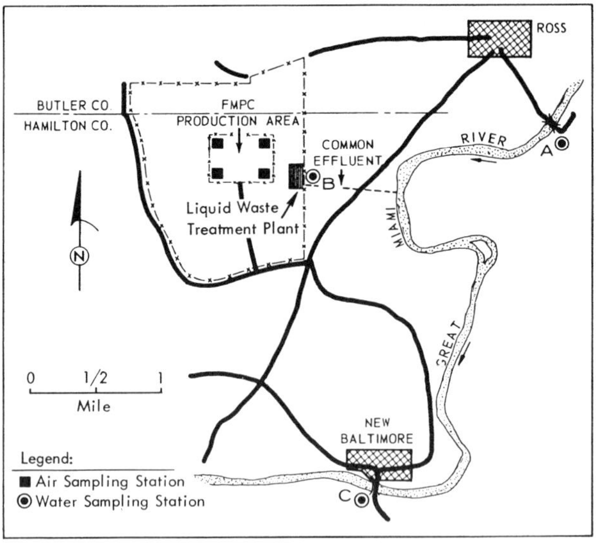

Figure 2. Air and water sampling stations, FMPC

Table 1. Radioactivity levels of airborne particulates, Feed Materials Production Center, July–December 1970

Location	Number of samples	Uranium concentration[a] (pCi/m³)			Alpha radioactivity[a] (pCi/m³)			Beta radioactivity[b] (pCi/m³)		
		Maximum	Minimum	Average	Maximum	Minimum	Average	Maximum	Minimum	Average
Onsite:										
Southwest	25	0.08	0.01	0.03	0.10	0.01	0.04	0.31	0.01	0.13
Northwest	25	.08	.01	.03	.16	.01	.04	.21	.04	.10
Northeast	25	.13	.01	.04	.18	.01	.05	.24	.02	.11
Southeast	25	.10	.01	.03	.16	.01	.04	.24	.01	.11
All onsite samples	100			0.03			0.04			0.11
Offsite:										
0–2 miles from FMPC	20	.11	.01	.05	.13	.01	.05	1.51	.05	.26
2–4 miles from FMPC	18	.08	.01	.03	.07	.01	.03	.21	.05	.13
4–8 miles from FMPC	28	.12	.01	.03	.12	.01	.03	.54	.01	.22
8–12 miles from FMPC	8	.04	.01	.03	.04	.01	.03	.15	.07	.10
All offsite samples	74			0.04			0.04			0.20

[a] AEC radiation protection standard—2 pCi/m³ (natural uranium).
[b] AEC radiation protection standard—1 nCi/m³ (thorium-234).

Water monitoring

Each of the individual production plants on the center has collection sumps and treatment equipment to remove the uranium from the process wastewater. The effluents from the plants are collected at a general sump for additional treatment and settling. The clear water from the sump is pumped to the river. The solid portion is pumped to a chemical waste pit for further settling. The clear effluent from the pit is then combined with three other types of project wastewater and discharged via a common effluent outfall into the Great Miami River. At location B, a water sampler is used in conjunction with a Parshal Flume to collect samples proportional to the flow of the combined effluent stream. These samples are removed and analyzed daily and results are used with measurements of river flow in calculating the radioactive contaminant concentrations added to the river. Weekly spot samples are obtained upstream (location A); a continuous sample is taken for a 24-hour period downstream (location C) and at least one sample is analyzed each week. All samples are analyzed for uranium, gross alpha and gross beta radioactivity, and radium-228, a daughter of thorium-232. Since radium-228 has the lowest AEC standard, control of this radionuclide and of the gross radioactivity insures that the AEC standards for the thorium decay chain are not exceeded.

The average concentrations of all sampled contaminants at the downstream position indicate that each contaminant was well below the AEC standard. It may be concluded from sampling and calculations that the FMPC effluent produced little change in the river's quality. The results of the FMPC water monitoring program for July–December 1970 are summarized in table 2.

Table 2. Radioactivity in the Great Miami River, Feed Materials Production Center, July–December 1970

Location	Number of samples	Uranium[a] (pCi/liter)			Alpha radioactivity[b] (pCi/liter)			Beta radioactivity[b] (pCi/liter)			Number of samples	Radium-228[c] (pCi/liter)		
		Maximum	Minimum	Average	Maximum	Minimum	Average	Maximum	Minimum	Average		Maximum	Minimum	Average
Sewer outfall[d] (location B)	183	7	<1	1	9	<1	1	24	<1	18	7	0.2	<0.1	0.1
Upstream from outfall (location A)	25	81	<1	7	105	<1	18	100	<1	23	6	.5	.5	.5
Downstream from outfall (location C)	27	16	<1	3	27	<1	7	27	<1	14	6	2.3	.5	1.0

[a] AEC standard—20 nCi/liter (natural uranium).
[b] AEC standard—3 nCi/liter (certain mixtures of alpha and beta emitters).
[c] AEC standard—30 pCi/liter (radium-228).
[d] Concentrations in the river as calculated from sewer outfall samples results.

2. Portsmouth Area Gaseous Diffusion Plant[3] July–December 1970

Goodyear Atomic Corporation
Piketon, Ohio

The separation of uranium isotopes by the gaseous diffusion process presents control problems similar to any chemical process using toxic solvents and extraction solutions. Natural uranium and thorium-234 are the most likely radionuclides to be released to the environment by the Portsmouth Area Gaseous Diffusion Plant. To verify the effectiveness of plant controls, environmental monitoring is conducted for evidence of alpha-particle, beta-particle, and gamma-ray emitters.

Air samples are taken at four offsite locations at a frequency of three 1,000 cubic-foot samples per week. In selecting sample locations, the prevailing wind direction and the locations with the highest radioactivity concentrations were considered.

Water samples are collected monthly at each location as shown in table 3. The one-half gallon sample is analyzed for gross alpha and gross beta-gamma radioactivity. Moderate volumes of low-level water wastes are discharged to the Scioto River system by way of Little Beaver Creek, Big Beaver Creek and Salt Creek. Sampling point distribution is as follows: one on each of the three plant drainage ditches; one on Salt Creek; three each on the Little Beaver Creek, the Big Beaver Creek, and the Scioto River.

[3] Data summarized from B. Kalmon and S. H. Hulett, "Environmental Radiation Levels and Concentrations, Second Half and Annual Summaries 1970," (April 20, 1970).

Table 3. Radioactivity in water, Portsmouth Plant, July–December 1970

Location	Number of samples	Alpha radioactivity concentration (pCi/m³)			Average as a percent of AEC standards[a]	Beta-gamma radioactivity concentration (pCi/m³)			Average as a percent of AEC standards[a]
		High	Low	Average		High	Low	Average	
1	6	31.5	1.8	7.6	0.03	7.0	7.0	7.0	0.04
2	6	36.0	.3	7.4	.02	18.0	7.0	8.6	.04
3[b]	5	216	.9	44.7	.15	194	7.0	36.8	.18
4	6	3.6	.3	1.9	.01	58.5	7.0	18.8	.09
5	6	185	.3	32.0	.11	49.5	7.0	22.1	.11
6	6	40.5	.5	11.7	.04	31.5	7.0	12.7	.06
7	6	148	50	78.8	.26	1,660	104	458	2.28
8	6	238	54	126	.42	1,400	166	560	2.80
10[b]	6	320	2.7	99	.33	81	7.0	25.9	.13
11[b]	6	356	99	194	.63	2,840	90	843	4.22
12	6	99	.5	19.9	.07	45	7.0	20.6	.10
13	6	63	2.3	28.9	.10	423	18.0	134	.67
14	6	9.0	.3	3.7	.01	27	7.0	10.1	.05
Summary	77	356	0.3	50.4	.17	2,840	7.0	166	0.83

[a] The AEC radiation protection standard for alpha radioactivity in water—30 nCi/liter; beta-gamma radioactivity—20 nCi/liter; sensitivity of analysis, alpha radioactivity—0.5 pCi/liter; beta radioactivity—14.0 pCi/liter.
[b] Composite sample.

Table 4. Radioactivity in air, Portsmouth Plant, July–December 1970

Location	Number of samples	Alpha radioactivity concentration (fCi/m³)			Average as a percent of AEC standards[a]	Beta-gamma radioactivity concentration (pCi/m³)			Average as a percent of AEC standards[a]
		High	Low	Average		High	Low	Average	
3	74	274	<16	48	1.21	1.4	<0.016	0.35	0.04
12	74	403	<16	32	.81	1.8	< .016	.35	.04
24	72	596	<16	48	1.21	1.9	< .016	.39	.04
29	72	386	<16	48	1.21	1.4	< .016	.31	.03
Summary	292	596	<16	44	1.10	1.9	<0.016	0.35	0.04

[a] The AEC radiation protection standard for alpha radioactivity in air—4 pCi/m³; beta-gamma radioactivity—1 nCi/m³; sensitivity of analysis for both alpha and beta radioactivity is 16.1 fCi/m³.

The frequency of composite sampling of water is determined by the flow at each of the three plant drainage ditches. The composite sample is collected in a 55-gallon drum. After thoroughly mixing the water in the drum, a one-half gallon sample is drawn off each month and the drum drained. The average concentrations in the Scioto River have been substantially below the AEC radiation protection standards for population groups in uncontrolled areas without need for control of releases.

For July–December 1970, the alpha and beta-gamma radioactivity in water, and the alpha radioactivity in air increased when compared with the first half of 1970.

Penetrating background dose rates for the second half of 1970 increased when compared with the first half of 1970. External gamma radiation levels are taken at each of four locations three times per week. Measurements are taken with a calibrated Geiger-Mueller tube at a height of 1 foot above ground level. These measurements are extrapolated to a distance of 3 feet. The 3-foot value has been experimentally determined as being approximately two-thirds that at the 1-foot level. These results are tabulated in mrem/h. Starting in January 1971, the background radiation level will be taken at 3 feet above ground.

Average alpha and beta-gamma radioactivity

Table 5 Background exposure rates, Portsmouth Plant, July–December 1970

Location	Background rate[a] (μrem/h)			Background average rate extrapolated to 3-foot level (μrem/h)
	High	Low	Average	
3	16.7	10.9	13.8	9.1
12	21.9	10.2	12.8	8.4
24	18.8	10.2	14.5	9.6
29	17.6	10.5	14.2	9.4
Summary	21.9	10.2	13.8	9.1

[a] Open-shield Geiger tube 1 foot above ground; limit of sensitivity—0.1 μrem/h.

concentrations in air are summarized in table 4. Table 3 contains the average alpha and beta-gamma radioactivity concentrations in water. The external gamma-ray levels, measured at the sampling locations shown in figure 3, are summarized in table 5. The overall average concentrations and background exposure rates for January–December 1969 and January–June 1970, July–December 1970, and January–December 1970 are presented in table 6.

Table 6. Comparison of average concentrations Portsmouth, January 1969–December 1970

Type of monitoring	Percent of AEC standard[a]			
	January–December 1969	January–June 1970	July–December 1970	January–December 1970
Air				
Alpha radioactivity	5.70	0.91	1.10	1.05
Beta-gamma radioactivity	.03	.07	.04	.05
Water				
Alpha radioactivity	.20	.08	.17	.12
Beta-gamma radioactivity	.32	.22	.83	.54
Background exposure extrapolated to 3 feet above ground level (μrem/h)	8.2	8.1	9.1	9.0

[a] See footnote (a) of table 3.

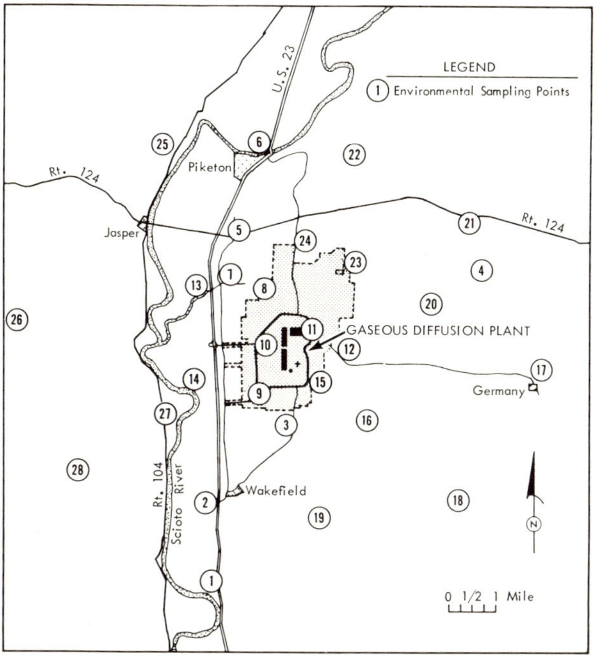

Figure 3. Sampling locations, Portsmouth Area Gaseous Diffusion Plant

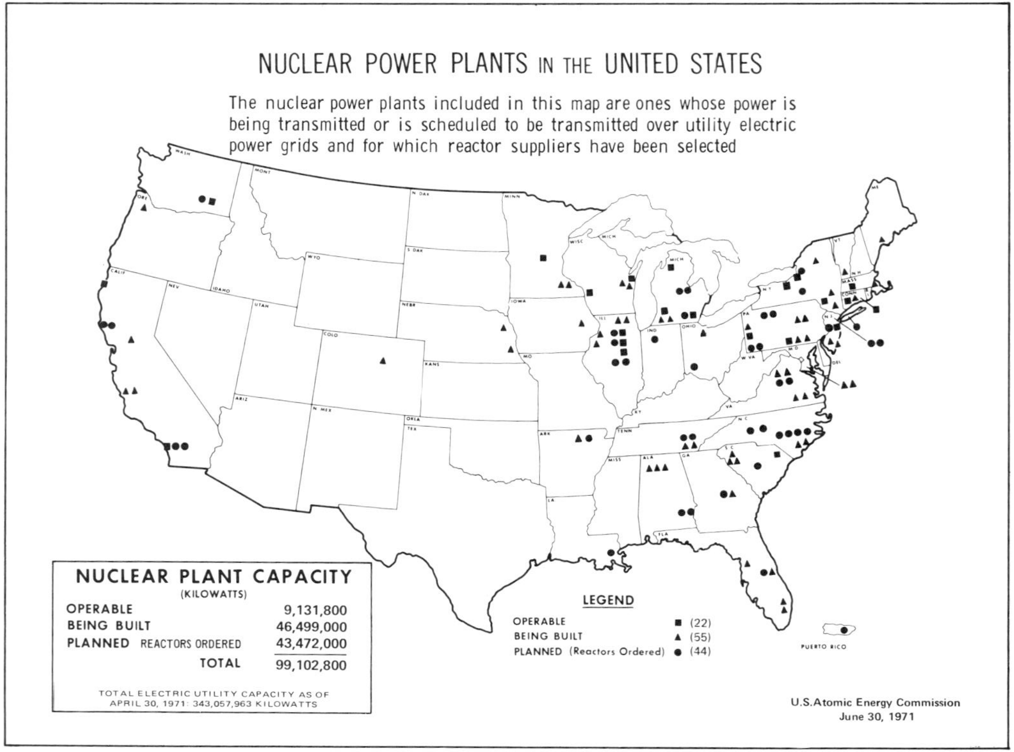

Figure 1. Nuclear power plants in the United States

SITE	PLANT NAME	CAPACITY (Net Kilowatts)	UTILITY	INITIAL DESIGN POWER
ALABAMA				
Decatur	Browns Ferry Nuclear Power Plant: Unit 1	1,065,000	Tennessee Valley Authority	1972
Decatur	Browns Ferry Nuclear Power Plant: Unit 2	1,065,000	Tennessee Valley Authority	1972
Decatur	Browns Ferry Nuclear Power Plant: Unit 3	1,065,000	Tennessee Valley Authority	1973
Dothan	Joseph M. Farley Nuclear Plant: Unit 1	829,000	Alabama Power Co.	1975
Dothan	Joseph M. Farley Nuclear Plant: Unit 2	829,000	Alabama Power Co.	1977
ARKANSAS				
London	Arkansas Nuclear One: Unit 1	820,000	Arkansas Power & Light Co.	1973
London	Arkansas Nuclear One: Unit 2	920,000	Arkansas Power & Light Co.	1975
CALIFORNIA				
Humboldt Bay	Humboldt Bay Power Plant: Unit 3	68,500	Pacific Gas & Electric Co.	1963
San Clemente	San Onofre Nuclear Generating Station: Unit 1	430,000	So. Calif. Ed. & San Diego Gas & El. Co.	1967
San Clemente	San Onofre Nuclear Generating Station: Unit 2	1,140,000	So. Calif. Ed. & San Diego Gas & El. Co.	1975
San Clemente	San Onofre Nuclear Generating Station: Unit 3	1,140,000	So. Calif. Ed. & San Diego Gas & El. Co.	1976
Diablo Canyon	Diablo Canyon Nuclear Power Plant: Unit 1	1,060,000	Pacific Gas & Electric Co.	1974
Diablo Canyon	Diablo Canyon Nuclear Power Plant: Unit 2	1,060,000	Pacific Gas & Electric Co.	1975
Clay Station	Rancho Seco Nuclear Generation Station	804,000	Sacramento Municipal Utility District	1973
Mendocino County	Mendocino Power Plant: Unit 1	1,128,000	Pacific Gas & Electric Co.	1976
Mendocino County	Mendocino Power Plant: Unit 2	1,128,000	Pacific Gas & Electric Co.	1978
COLORADO				
Platteville	Ft. St. Vrain Nuclear Generating Station	330,000	Public Service Co. of Colorado	1972
CONNECTICUT				
Haddam Neck	Haddam Neck Plant	575,000	Conn. Yankee Atomic Power Co.	1967
Waterford	Millstone Nuclear Power Station: Unit 1	652,100	Northeast Utilities	1970
Waterford	Millstone Nuclear Power Station: Unit 2	828,000	Northeast Utilities	1974

Figure 1. Nuclear power plants in the United States—continued

SITE	PLANT NAME	CAPACITY (Net Kilowatts)	UTILITY	INITIAL DESIGN POWER
FLORIDA				
Turkey Point	Turkey Point Station: Unit 3	693,000	Florida Power & Light Co.	1971
Turkey Point	Turkey Point Station: Unit 4	693,000	Florida Power & Light Co.	1972
Red Level	Crystal River Plant: Unit 3	858,000	Florida Power Corp.	1972
Ft. Pierce	Hutchinson Island: Unit 1	800,000	Florida Power and Light Co.	1974
Ft. Pierce	Hutchinson Island: Unit 2	800,000	Florida Power and Light Co.	1976
GEORGIA				
Baxley	Edwin I. Hatch Nuclear Plant: Unit 1	786,000	Georgia Power Co.	1973
Baxley	Edwin I. Hatch Nuclear Plant: Unit 2	786,000	Georgia Power Co.	1976
ILLINOIS				
Morris	Dresden Nuclear Power Station: Unit 1	200,000	Commonwealth Edison Co.	1960
Morris	Dresden Nuclear Power Station: Unit 2	809,000	Commonwealth Edison Co.	1970
Morris	Dresden Nuclear Power Station: Unit 3	809,000	Commonwealth Edison Co.	1971
Zion	Zion Nuclear Plant: Unit 1	1,050,000	Commonwealth Edison Co.	1972
Zion	Zion Nuclear Plant: Unit 2	1,050,000	Commonwealth Edison Co.	1973
Cordova	Quad-Cities Station: Unit 1	809,000	Comm. Ed. Co.-Ia.-Ill. Gas & Elec. Co.	1971
Cordova	Quad-Cities Station: Unit 2	809,000	Comm. Ed. Co.-Ia.-Ill. Gas & Elec. Co.	1972
Seneca	LaSalle Co. Nuclear Station: Unit 1	1,078,000	Comm. Ed. Co.-Ia.	1975
Seneca	LaSalle Co. Nuclear Station: Unit 2	1,078,000	Comm. Ed. Co.-Ia.	1976
*	—	1,100,000	Comm. Edison Co.	1978
*	—	1,100,000	Comm. Edison Co.	1979
INDIANA				
Dune Acres	Bailly Generating Station	660,000	Northern Indiana Public Service Co.	1976
IOWA				
Cedar Rapids	Duane Arnold Energy Center: Unit 1	529,700	Iowa Electric Light and Power Co.	1973
LOUISIANA				
Taft	Waterford Generating Station: Unit 1	1,165,000	Louisiana Power & Light Co.	1976
MAINE				
Wiscasset	Maine Yankee Atomic Power Plant	790,000	Maine Yankee Atomic Power Co.	1972
MARYLAND				
Lusby	Calvert Cliffs Nuclear Power Plant: Unit 1	845,000	Baltimore Gas and Electric Co.	1973
Lusby	Calvert Cliffs Nuclear Power Plant: Unit 2	845,000	Baltimore Gas and Electric Co.	1974
MASSACHUSETTS				
Rowe	Yankee Nuclear Power Station	175,000	Yankee Atomic Electric Co.	1961
Plymouth	Pilgrim Station	655,000	Boston Edison Co.	1971
MICHIGAN				
Big Rock Point	Big Rock Point Nuclear Plant	70,300	Consumers Power Co.	1963
South Haven	Palisades Nuclear Power Station	700,000	Consumers Power Co.	1971
Lagoona Beach	Enrico Fermi Atomic Power Plant: Unit 1	60,900	Detroit Edison Co.	1970
Lagoona Beach	Enrico Fermi Atomic Power Plant: Unit 2	1,123,000	Detroit Edison Co.	1973
Bridgman	Donald C. Cook Plant: Unit 1	1,054,000	Indiana & Michigan Electric Co.	1973
Bridgman	Donald C. Cook Plant: Unit 2	1,060,000	Indiana & Michigan Electric Co.	1974
Midland	Midland Nuclear Power Plant: Unit 1	492,000	Consumers Power Co.	1976
Midland	Midland Nuclear Power Plant: Unit 2	818,000	Consumers Power Co.	1977
MINNESOTA				
Monticello	Monticello Nuclear Generating Plant	545,000	Northern States Power Co.	1971
Red Wing	Prairie Island Nuclear Generating Plant: Unit 1	530,000	Northern States Power Co.	1972
Red Wing	Prairie Island Nuclear Generating Plant: Unit 2	530,000	Northern States Power Co.	1974

NEBRASKA				
Fort Calhoun	Ft. Calhoun Station: Unit 1	457,400	Omaha Public Power District	1972
Brownville	Cooper Nuclear Station	778,000	Nebraska Public Power District and Iowa Power and Light Co.	1973
NEW JERSEY				
Toms River	Oyster Creek Nuclear Power Plant: Unit 1	560,000	Jersey Central Power & Light Co.	1969
Lacey Township	Forked River Generating Station: Unit 1	1,140,000	Jersey Central Power & Light Co.	1975
Salem	Salem Nuclear Generating Station: Unit 1	1,050,000	Public Service Electric and Gas, N. J.	1973
Salem	Salem Nuclear Generating Station: Unit 2	1,050,000	Public Service Electric and Gas, N. J.	1974
Bordentown	Newbold Nuclear Generating Station: Unit 1	1,088,000	Public Service Electric and Gas, N. J.	1975
Bordentown	Newbold Nuclear Generating Station: Unit 2	1,088,000	Public Service Electric and Gas, N. J.	1977
NEW YORK				
Indian Point	Indian Point Station: Unit 1	265,000	Consolidated Edison Co.	1963
Indian Point	Indian Point Station: Unit 2	873,000	Consolidated Edison Co.	1971
Indian Point	Indian Point Station: Unit 3	965,000	Consolidated Edison Co.	1973
Scriba	Nine Mile Point Nuclear Station	625,000	Niagara Mohawk Power Co.	1970
Rochester	R. E. Ginna Nuclear Power Plant: Unit 1	420,000	Rochester Gas & Electric Co.	1970
Brookhaven	Shoreham Nuclear Power Station	819,000	Long Island Lighting Co.	1975
Lansing	Bell Station	838,000	New York State Electric & Gas Co.	1978
Verplanck	Verplanck: Unit 1	1,115,000	Consolidated Edison Co.	1977
Scriba	James A. Fitzpatrick Nuclear Power Plant	821,000	Power Authority of State of N. Y.	1973
NORTH CAROLINA				
Southport	Brunswick Steam Electric Plant: Unit 1	821,000	Carolina Power and Light Co.	1975
Southport	Brunswick Steam Electric Plant: Unit 2	821,000	Carolina Power and Light Co.	1974
Cowans Ford Dam	Wm. B. McGuire Nuclear Station: Unit 1	1,150,000	Duke Power Co.	1975
Cowans Ford Dam	Wm. B. McGuire Nuclear Station: Unit 2	1,150,000	Duke Power Co.	1977
Bonsal	Shearon Harris Plant: Unit 1	900,000	Carolina Power & Light Co.	1977
Bonsal	Shearon Harris Plant: Unit 2	900,000	Carolina Power & Light Co.	1978
Bonsal	Shearon Harris Plant: Unit 3	900,000	Carolina Power & Light Co.	1979
Bonsal	Shearon Harris Plant: Unit 4	900,000	Carolina Power & Light Co.	1980
OHIO				
Oak Harbor	Davis-Besse Nuclear Power Station	872,000	Toledo Edison-Cleveland Electric Illuminating Co.	1974
Moscow	Wm. H. Zimmer Nuclear Power Station: Unit 1	810,000	Cincinnati Gas & Electric Co.	1974
OREGON				
Prescott	Trojan Station	1,130,000	Portland General Electric Co.	1974
PENNSYLVANIA				
Peach Bottom	Peach Bottom Atomic Power Station: Unit 1	40,000	Philadelphia Electric Co.	1967
Peach Bottom	Peach Bottom Atomic Power Station: Unit 2	1,065,000	Philadelphia Electric Co.	1973
Peach Bottom	Peach Bottom Atomic Power Station: Unit 3	1,065,000	Philadelphia Electric Co.	1974
Pottstown	Limerick Generating Station: Unit 1	1,065,000	Philadelphia Electric Co.	1975
Pottstown	Limerick Generating Station: Unit 2	1,065,000	Philadelphia Electric Co.	1977
Shippingport	Shippingport Atomic Power Station: Unit 1	90,000	Duquesne Light Co.	1957
Shippingport	Beaver Valley Power Station: Unit 1	847,000	Duquesne Light Co.-Ohio Edison Co.	1973
Middletown	Three Mile Island Nuclear Station: Unit 1	831,000	Metropolitan Edison Co.	1972
Middletown	Three Mile Island Nuclear Station: Unit 2	907,000	Jersey Central Power & Light Co.	1974
Berwick	Susquehanna Steam Electric Station: Unit 1	1,052,000	Pennsylvania Power and Light	1977
Berwick	Susquehanna Steam Electric Station: Unit 2	1,052,000	Pennsylvania Power and Light	1980

Figure 1. Nuclear power plants in the United States—continued

SITE	PLANT NAME	CAPACITY (Net Kilowatts)	UTILITY	INITIAL DESIGN POWER
SOUTH CAROLINA				
Hartsville	H. B. Robinson S. E. Plant: Unit 2	700,000	Carolina Power & Light Co.	1971
Seneca	Oconne Nuclear Station: Unit 1	841,000	Duke Power Co.	1971
Seneca	Oconee Nuclear Station: Unit 2	886,000	Duke Power Co.	1972
Seneca	Oconee Nuclear Station: Unit 3	886,000	Duke Power Co.	1973
Parr	Virgil C.Summer Nuclear Station: Unit 1	900,000	South Carolina Electric & Gas Co.	1977
TENNESSEE				
Daisy	Sequoyah Nuclear Power Plant: Unit 1	1,124,000	Tennessee Valley Authority	1974
Daisy	Sequoyah Nuclear Power Plant: Unit 2	1,124,000	Tennessee Valley Authority	1974
Spring City	Watts Bar Nuclear Plant: Unit 1	1,169,000	Tennessee Valley Authority	1976
Spring City	Watts Bar Nuclear Plant: Unit 2	1,169,000	Tennessee Valley Authority	1977
VERMONT				
Vernon	Vermont Yankee Generating Station	513,900	Vermont Yankee Nuclear Power Corp.	1971
VIRGINIA				
Gravel Neck	Surry Power Station: Unit 1	780,000	Virginia Electric & Power Co.	1972
Gravel Neck	Surry Power Station: Unit 2	780,000	Virginia Electric & Power Co.	1972
Mineral	North Anna Power Station: Unit 1	845,000	Virginia Electric & Power Co.	1974
Mineral	North Anna Power Station: Unit 2	845,000	Virginia Electric & Power Co.	1975
Mineral	North Anna Power Station: Unit 3	900,000	Virginia Electric & Power Co.	1977
Mineral	North Anna Power Station: Unit 4	900,000	Virginia Electric & Power Co.	1978
WASHINGTON				
Richland	N-Reactor/WPPSS Steam	790,000	Atomic Energy Commission	1966
Richland	Hanford No. 2	1,103,000	Washington Public Power Supply System	1977
WISCONSIN				
Genoa	Genoa Nuclear Generating Station	50,000	Dairyland Power Cooperative	1969
Two Creeks	Point Beach Nuclear Plant: Unit 1	497,000	Wisconsin Michigan Power Co.	1971
Two Creeks	Point Beach Nuclear Plant: Unit 2	497,000	Wisconsin Michigan Power Co.	1971
Carlton	Kewaunee Nuclear Power Plant: Unit 1	540,000	Wisconsin Public Service Co.	1972
PUERTO RICO				
Puerto De Jobas	Aguirre Nuclear Power Plant	583,000	Puerto Rico Water Resources Authority	1975
*	—	1,201,000	Tennessee Valley Authority	1977
*	—	1,201,000	Tennessee Valley Authority	1978

* Site not selected.

Figure 1. Nuclear power plants in the United States—continued

Environmental Levels of Radioactivity at Atomic Energy Commission Installations

1. Atomics International[1] January–June 1970

North American Rockwell Corporation
Canoga Park, California

Atomics International, a division of North American Rockwell Corporation, has engaged in atomic energy research and development since 1946. The company designs, develops, and constructs nuclear reactors for central station and compact power plants for medical, industrial, and scientific applications.

The company headquarters is located in Canoga Park, Calif., approximately 23 miles northwest of downtown Los Angeles. The 290-acre Nuclear Development Field Laboratory (Santa Susana Facility), equipped with extensive testing facilities for the support of advanced nuclear studies, is located in the Simi Hills of Ventura County, approximately 29 miles northwest of downtown

---

[1] Summarized from "Environmental Monitoring, Semi-annual Report, January 1, 1970 to June 30, 1970" Atomics International, Division of North American Rockwell Corporation.

Los Angeles. The location of the above sites in relation to nearby communities is shown in figure 1.

The basic concept of radiological hazard control at Atomics International requires adequate containment of radioactive materials and, through rigid operational controls, minimizes effluent releases and external radiation levels. The environmental monitoring program provides a measure of the effectiveness of the company's radiological safety procedures and of engineering safeguards incorporated into facility designs.

The onsite environs of Atomics International headquarters and Nuclear Development Field Laboratory (NDFL) are surveyed monthly to determine the concentration of radioactivity in typical surface soil, vegetation, and water samples. The offsite environs are sampled monthly, except for soil and vegetation which are sampled quarterly. In addition, continuous environmental air monitoring at the sites provides information concerning long-lived airborne particulate radioactivity.

Air monitoring

Environmental air sampling is conducted continuously at the headquarters and NDFL sites with automatic air samplers operating on 24-hour sampling cycles. Airborne particulate radioactivity is collected on HV-70 filter paper which is automatically changed at the end of each sampling period. The filter is removed from the sampler and counted after the radioactivity is allowed to decay for at least 72 hours. The volume of a typical daily environmental air sample is approximately 20 cubic meters. The average concentration of long-lived beta-gamma radioactivity on airborne particulates is presented in table 1 for January–June 1970.

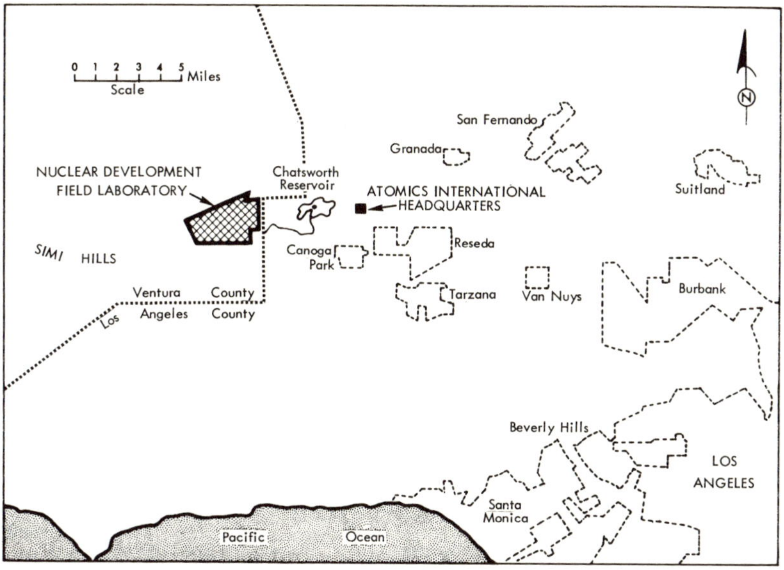

Figure 1. Atomics International facilities and vicinity

Table 1. Beta-gamma radioactivity of airborne particulates, Atomics International, January–June 1970

Location	Number of samples	Average concentration[a] (pCi/m³)
Headquarters	324	0.46
NDFL	1,212	.46

[a] Minimum detectable level—0.04 pCi/m³.

When abnormally high airborne radioactivities are observed, the radioactivity decay data are plotted to determine the presence of long-lived isotopes other than naturally occurring radon, thoron, and daughters. If fallout is suspected, the decay characteristics are observed. If the radioactivity decays as a function of $t^{-1.2}$, the data curve is extrapolated in order to determine the date of origin. These data are compared with the dates of publicized nuclear detonations to determine if the abnormal airborne radioactivity was caused by such detonations.

A graph of averaged long-lived airborne radioactivity concentrations detected at the Headquarters and NDFL facilities during the first half of 1970 is presented in figure 2. The graph shows a generally increasing trend through the spring months with a prominent peak noted in May.

Water monitoring

Process water used at the NDFL is obtained from Ventura County Water District No. 10 and distributed onsite by the same piping system previously used when process water was supplied by onsite wells. Pressure is provided by elevated storage tanks, one 50,000-gallon and one 500,000-gallon tank onsite. While clinically potable, the water is not used for drinking. Bottled potable water is delivered by a vendor and is not analyzed. Water from the pipe system is sampled monthly at two locations. The average process water radioactivity concentration is presented in table 2.

Soil, vegetation, and water are sampled monthly at Chatsworth Reservoir which is operated by the Los Angeles City Department of Water and Power. Normally, one water sample is obtained from the lake surface and a second sample is obtained from

Table 2. Process water radioactivity, NDFL site January–June 1970

Type of radioactivity	Number of samples	Average concentration (pCi/liter)
Alpha	12	0.05
Beta-gamma	12	5.6

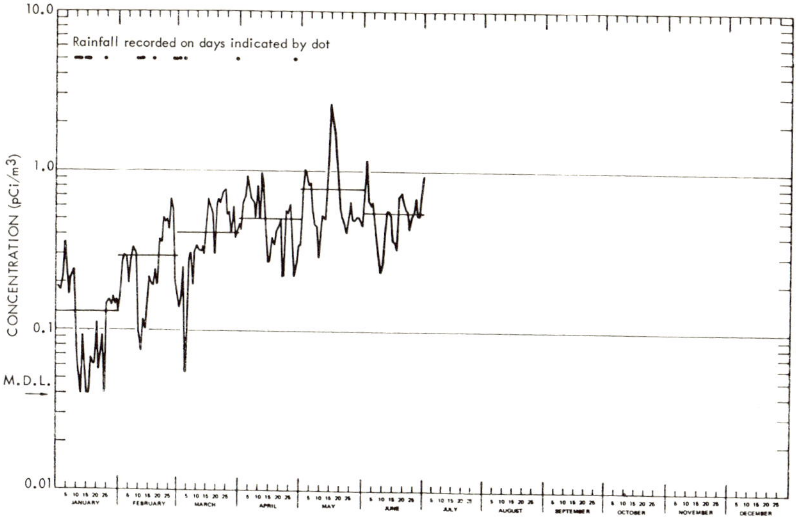

Figure 2. Long-lived airborne radioactivity at Atomics International, 1970

the reservoir water supply inlet located on the north side of the lake. The lake was drained in July 1969 for construction, thereby precluding sampling for most of the last half of 1969, and the current reporting period. The average radioactivity for both surface and supply water samples is presented in table 3.

Table 3. Chatsworth Reservoir water radioactivity, Atomics International, January–June 1970

Sample[a]	Type of radioactivity	Number of samples	Average concentration (pCi/liter)
Supply inlet	Alpha	6	0.18
	Beta-gamma	6	5.9

[a] No samples were collected from the lake surface for this period.

Surface discharged waters from NDFL facilities drain into holding reservoirs on adjacent property. When full, the main reservoir is drained into Bell Creek, a tributary of the Los Angeles River in the San Fernando Valley, Los Angeles County. Pursuant to the requirements of Los Angeles Regional

Water Quality Control Board Resolution 66-49 of September 21, 1966, an environmental sampling station has been established in Bell Creek Canyon, approximately 2.5 miles downstream from the south NDFL boundary. Samples, obtained and analyzed monthly, include stream bed mud, vegetation, and water. Average radioactivity concentrations in the main holding reservoir and Bell Creek samples are presented in table 4.

Table 4. Radioactivity in the Rocketdyne reservoir and Bell Creek,[a] January–June 1970

Sample description (units)	Number of samples	Alpha radioactivity	Beta radioactivity
Reservoir station 6, water[a] (pCi/liter)	6	0.05	6.7
Reservoir station 12, water[a] (pCi/liter)	6	.07	7.6
Bell Creek, mud (pCi/g)	6	.34	25
Bell Creek, vegetation (pCi/g ash)	6	.29	171
Bell Creek, water (pCi/liter)	6	.019	4.0

[a] Location not shown on figure 1.

Soil and vegetation monitoring

Soil and vegetation are regularly sampled at 25 locations. Eleven sampling stations are located within the boundaries of Atomics International's sites and are referred to as "onsite" stations. The remaining 14 stations, located within a 10-mile radius of sites, are referred to as "offsite" stations.

Surface soil types available for sampling range from decomposed granite to clay and loam. Samples are taken from the top half-inch layer of ground surface. The soil samples are packaged and sealed in plastic containers and returned to the laboratory for analysis. Radioactivity in soil samples is presented in table 5.

Table 5. Radioactivity in the soil, Atomics International, January–June 1970

Area	Type of radioactivity	Number of samples	Average concentration (pCi/g)
Onsite	Alpha	72	0.39
	Beta-gamma	72	28
Offsite	Alpha	24	.41
	Beta-gamma	24	26

Vegetation samples obtained in the field are of the same plant type wherever possible, generally, sunflower or wild tobacco plant leaves. These types maintain a more active growth rate during the dry season than do most natural vegetation indigenous to the local area. Vegetation leaves are stripped from plants and transferred to the laboratory for analysis. Plant root systems are not routinely sampled. Radioactivity in vegetation samples is presented in table 6.

Table 6. Radioactivity in vegetation, Atomics International, January–June 1970

Area	Type of radioactivity	Number of samples	Average concentration (pCi/g ash)
Onsite	Alpha	72	0.41
	Beta-gamma	72	169
Offsite	Alpha	24	.30
	Beta-gamma	24	154

2. Neutron Devices Department[2]
January–June 1970

General Electric Company
St. Petersburg, Fla.

The Neutron Devices Department (figure 3) is an electronic component production facility. The plant maintains an environmental monitoring program to measure the levels of radioactive environmental contamination associated with plant effluents. These measurements serve as an index of the effectiveness of the plant's contamination control measures. Sewer effluent, surface water, milk and air within 6 miles of the plant are monitored for radioactivity. The radioactivity concentrations, weighted by atmospheric and stream dilution factors, are compared against the AEC standards for continuous nonoccupational exposure as presented in AEC Manual chapter 0524.

Sewer effluent monitoring

A combined sewer effluent sample is obtained daily near the perimeter of the plant's property. During the sampling period, 20 of 137 samples analyzed showed detectable concentrations of tritium (detectable level-90 nCi/liter). The maximum concentrations (1.10 μCi/liter, detected on January 20, 1970) represented 37 percent of the continuous nonoccupational exposure guide. Calculations based on radioactivity releases from the process waste system and the plant's water discharges indicate that the average tritium concentration in the combined sewer effluent for the first half of 1970 was less than 7.9 percent of the AEC standard for continuous nonoccupational exposure.

[2] Summarized from "Environmental Monitoring, January 1 through June 30, 1970," General Electric Company, Neutron Devices Department, St. Petersburg, Fla.

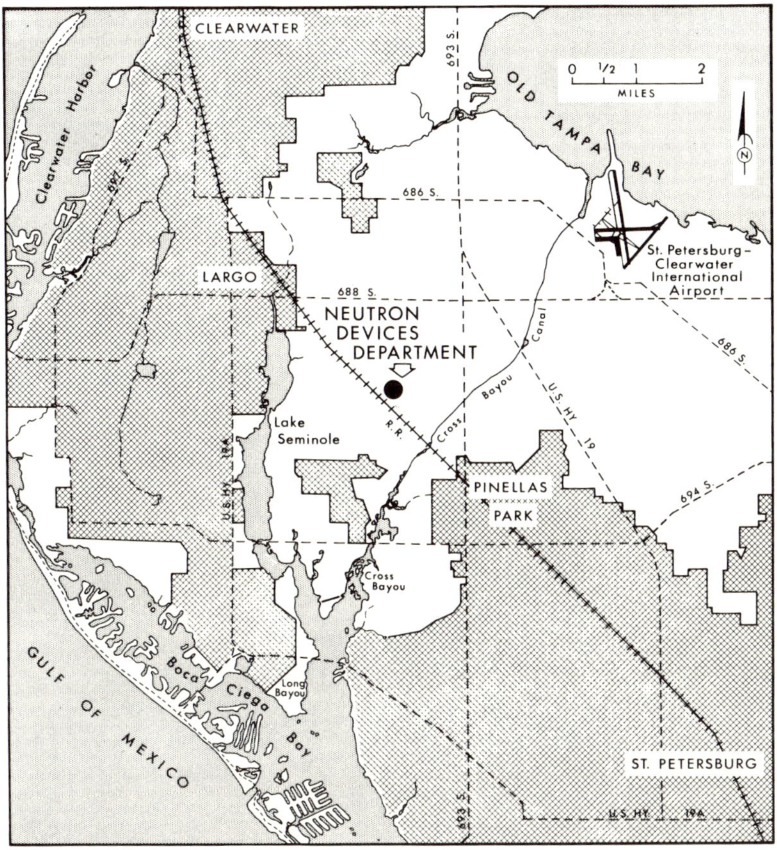

Figure 3. Location of the Neutron Devices Department

Surface water sampling

Surface water samples are collected at monthly intervals at selected locations within 6 miles of the plant. The sampling areas are determined by interrelating the concentrations of radioactivity in exhaust stack effluent with meteorological data. There were no indications of tritium (detectable level-50 nCi/liter) in the 137 surface water samples analyzed during the sampling period.

Milk sampling results

Analyses of 14 milk samples, collected from one local dairy farm by the Pinellas County Health Department, revealed no detectable concentrations (detectable level-90 nCi/liter) of tritium.

Air sampling results

One tritium gas and three tritium oxide in air samples obtained at the perimeter of the plant property, downwind from the exhaust stack revealed no detectable amounts of tritium gas (detectable level-20 $\mu Ci/m^3$) or tritium oxide (detectable level-5 nCi/m^3).

Summary

Offsite radioactivity concentrations encountered by the general population are substantially lower than the AEC standards for continuous nonoccupational exposure.

Environmental Levels of Radioactivity at Atomic Energy Commission Installations

1. Lawrence Radiation Laboratory [1] July–December 1970

University of California
Berkeley, Calif.

Berkeley site

The Berkeley site of the Lawrence Radiation Laboratory (LRL) is situated on the western slope of the range of hills running along the eastern side of San Francisco Bay. The laboratory area is largely separate from the main campus of the University (figure 1).

[1] Summarized from "Results of Environmental Radioactivity Sampling Program, July–December 1970," Lawrence Radiation Laboratory, Livermore and Berkeley, Calif. 94720.

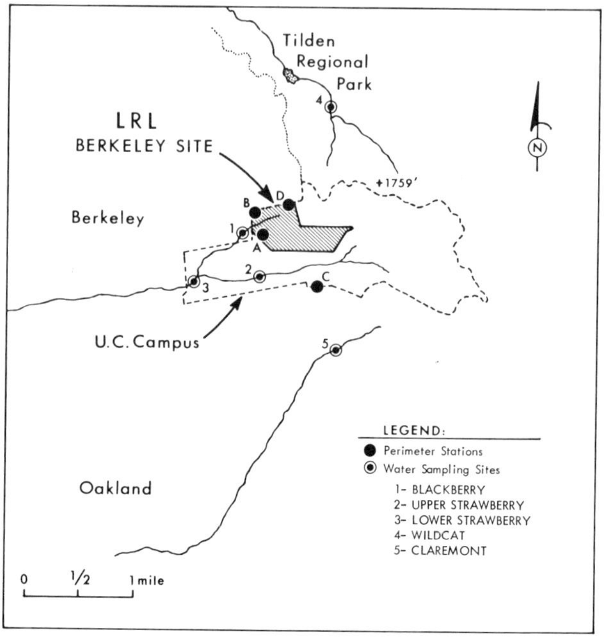

Figure 1. Environmental sampling locations at the Berkeley site, Lawrence Radiation Laboratory

To the north and south of the laboratory area are residential areas of the cities of Berkeley and Oakland. The Berkeley campus of the University of California is on the west; to the east, are the Lawrence Hall of Science and the Space Sciences Laboratory, beyond them lies uninhabited watershed land and the Tilden Regional Park.

The most prevalent wind direction is westerly. During the summer months, the afternoon sea breeze establishes a very pronounced prevailing westerly wind direction. At other times, the direction is less predictable. Annual rainfall is 23 inches, almost all of which falls between November

1 and May 1. The prevailing wind direction during precipitation is southerly.

The laboratory carries on a wide ranging program of general research in the fields of both physical and biological sciences. Facilities include a number of large accelerators, and various physics, chemistry, biology, and medical research laboratories.

The basic policy at LRL has long been to prevent, as far as possible, any release of radioactive material to the environment, no matter how small. No deliberate releases are sanctioned, except where no practical method for containment has yet been developed, and quantities are small compared with standards established by the International Commission on Radiological Protection (ICRP) and the Atomic Energy Commission (AEC).

The most sensitive measurements of releases are those taken closest to the source itself, before dilution makes detection of small quantities more difficult. The stack sampling program, therefore, provides the most useful information for controlling releases to the atmosphere. Over 100 separate exhausts from hoods and glove-box manifolds are sampled. The total quantity released from these stacks during July–December 1970 was 300 nCi of alpha radioactivity, 420 mCi of carbon-14, 18.3 curies of tritium and 53 μCi of other beta-particle emitters. The average concentrations in the total exhaust, before any dilution by the atmosphere, was 25 percent of the maximum permissible concentrations for offsite breathing zone air as given by the ICRP.

In addition to the careful sampling of stacks, an environmental air sampling program is also carried on to make sure no undetected releases occur. These samples provide a direct measurement of possible exposure to the nearby population. The sampling stations designated "local area" are scattered around the site and provide samples of the atmosphere on the site itself. The "perimeter" samples are taken at the boundary line of University property, in the direction of populated areas. From the results of these samples, there appears to have been no significant exposure from radioactive materials released by LRL. The levels of radioactivity observed in each type of sample are presented in table 1.

At each of these environmental stations, rain or dry fallout is also collected. Open 15-inch-diameter polyethylene bags form the collecting vessel. If no

Table 1. Atmospheric monitoring, LRL Berkeley site July–December 1970

Sampling locations (number of locations)	Number of samples	Average concentration (pCi/m³)			
		Alpha radioactivity		Beta radioactivity	
		Average	Maximum	Average	Maximum
Local area (10)	258	0.000	<0.005	0.16	0.50
Perimeter					
A:	26	.000	.005	.15	.48
B:	23	.000	< .005	.16	.40
C:	26	.000	< .005	.16	.44
D:	24	.000	< .005	.19	.41
Summary	99	.000	< .005	.16	.48

rain has fallen, the bags are rinsed out with water to obtain a sample. Since on the average, the more remote perimeter stations show about the same radioactivity as the nearer ones, the radioactivity does not result from laboratory operations. The radioactivity levels in samples collected at each location are summarized in table 2.

Table 2. Total deposition, LRL Berkeley site July–December 1970

Sampling locations (number of locations)	Number of samples	Average deposition (nCi/m²)			
		Alpha radioactivity		Beta radioactivity	
		Average	Maximum	Average	Maximum
Local area (10)	54	0.008	0.046	2.99	4.01
Perimeter (4)	24	.000	.019	3.73	4.86

Table 3. Summary of sewer data, LRL Berkeley site, July–December 1970

Sewer	Total quantities discharged 6 months			Net concentrations (pCi/liter)				
	Total volume (M1)	Total alpha radioactivity (μCi)	Total beta radioactivity (mCi)	Number of samples	Alpha		Beta	
					Average	Maximum	Average	Maximum
Hearst	119.3	11.3	4.30	24	0.09	3.2	36.1	185
Strawberry	38.4	2.4	.49	24	.06	1.8	12.7	25
Summary	157.7	13.7	4.79		0.09		30.3	

All liquid waste known to be radioactive is collected, solidified, and shipped away. Other liquid wastes are discharged directly to the municipal sewer system. Occasionally, small quantities of radioisotopes are accidentally released to the sewer system. There are two outfalls, each of which is monitored by a continuous proportional sampling system to insure that no significant quantities have been discharged (table 3). The total concentration (alpha-plus-beta radioactivity) in sewage is only 31 percent of the standard for drinking water and therefore does not constitute an exposure threat.

The storm drainage from the laboratory flows into the surface stream system. These surface streams are exposed as they run through the University property and are sampled at three places.

Results are listed as "onsite streams" in table 4. Two nearby offsite streams are also sampled to provide a comparison. All concentrations are less than the standard for drinking water. The results from the water sampling program are presented in table 4.

Table 4. Water monitoring, LRL Berkeley site July–December 1970

Type and source of sample	Number of samples	Concentration (pCi/liter) Alpha radioactivity Average	Alpha radioactivity Maximum	Beta radioactivity Average	Beta radioactivity Maximum
Surface water:					
Onsite streams					
Blackberry	26	0.30	1.2	3.27	14.3
Lower Strawberry	26	.33	8.5	7.37	29.3
Upper Strawberry	26	.14	1.6	2.95	9.3
Summary		0.26	8.5	4.53	29.3
Offsite streams:					
Claremont	26	.68	2.9	3.00	13.3
Wildcat	26	.16	1.7	2.19	5.0
Summary		0.42	2.9	2.60	13.3
Tap water	24	.005	.37	1.35	3.2

Neutron and gamma-ray fields are measured at each of the four perimeter stations as well as at several locations within the site. Background measurements are obtained from a leased building used as a low-level counting room in downtown Berkeley. This building is remote from the main LRL site and measurements are not influenced by accelerator operations.

At each station, gamma measurements are made by a recording G-M detector, standardized with a calibrated radium source. These instruments operate unattended and each can operate for 1 week on its internal battery should external power fail.

Several neutron detectors are employed at each station. The primary detector is a BF_3 gas proportional counter in a 2½ inch-thick paraffin-lined moderator. This detector is sensitive in the region from 0.1 to 20 MeV, where 75 percent of the biologically significant neutrons exist. A polyethylene-lined proton recoil neutron detector provides a measure of the energy flux density. The ratio of energy flux density to neutron flux density gives the average neutron energy which allows conversion of the data to dose in rems. Cobalt disks in moderators are maintained as long term integrating neutron monitors at many sites throughout the laboratory as well as at each perimeter station.

Cosmic ray intensity is continuously monitored by a meson detector with an effective angle of acceptance of 2.3 steradians (in the upward direction). This provides information on the background neutron levels contributed by cosmic rays.

At each station, total radiation levels are well below the standards set for the general public by the ICRP and National Council on Radiation Protection and Measurements (NCRP). Results are presented in table 5.

Table 5. Beta-gamma and fast neutron surveillance results, Berkeley site, July–December 1970

Station	Total beta-gamma exposure (mrem) (background subtracted)	Total fast neutron exposure[a] (mrem) (background subtracted)
A	11.9	54.5
B	16.0	28.4
C	Background	Background
D	3.0	22.5
Downtown Berkeley	(b)	(b)

[a] $2.45 \times 10^7 n/cm^2 = 1$ rem.
[b] Detector not in service.

Livermore site

The Livermore site of LRL (figure 2) is located in the Livermore Valley midway between the Pacific coast and the San Joaquin Valley in

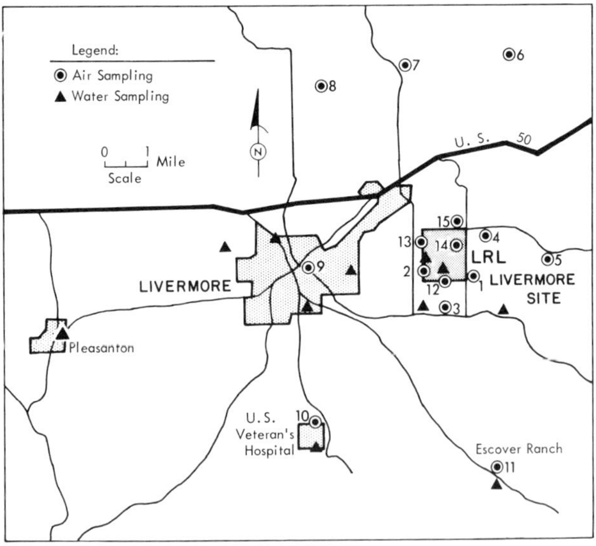

Figure 2. Environmental sampling locations at the Livermore site, Lawrence Radiation Laboratory

northern California, approximately 50 miles southeast of San Francisco. Shielded from the ocean by the western hills, the Livermore Valley has a warm, dry climate. Annual rainfall is about 14 inches with most of the precipitation occurring during the winter months. Prevailing winds are from the west, but inversions are frequent during the night.

Agriculture is the principal activity in the Livermore Valley. Roses, grain, hay, and grapes are the major products. Several cattle and sheep ranches surround the Livermore site.

The Livermore site comprises an area of 1-square mile, approximately 3 miles east of the city of Livermore. Livermore and Pleasanton, with a combined population of 45,000 are populated areas of primary interest to the Livermore site.

An environmental sampling program is maintained to provide information regarding the effectiveness of control measures and to determine whether any radiological changes in the environment are the result of laboratory operations. The sampling program includes air particulates, soil, domestic water, sewer effluent, sewage plant products, milk, and vegetation. The milk samples are obtained from one dairy in the Livermore Valley and from another dairy near Site 300.

Techniques

The LRL Environmental Program is currently undergoing a significant upgrading. This upgrading will greatly improve the analytical sensitivities in order to determine actual background levels of specific radionuclides in the environment. During the transition period, the program will consist of two parts, (1) the gross radioactivity measurements similar to those made in the past, and (2) the special measurements designed to measure the activity of specific radionuclides in environmental samples at background levels.

The analytical techniques used in the above measurements are provided below:

1. Gross radioactivity measurements:

An automated counting system is used for all environmental air samples. This system utilizes gas-flow proportional detectors for both alpha- and beta-particle counting. The alpha-particle detector is approximately 20 percent efficient for the 5.1 MeV alpha of plutonium-239. The efficiency for the 0.54 MeV and 2.27 MeV betas of strontium-90 in equilibrium with yttrium-90 is about 28 percent.

Soil, water, and sewage samples are counted on an automatic low-background system that detects alpha particles with an efficiency of about 34 percent at 5.1 MeV and strontium-yttrium-90 beta particles with an efficiency of approximately 50 percent.

The milk samples are concentrated using a vacuum-operated, foam-entrainment evaporator so that about 9 liters of whole milk are condensed to 2 liters by evaporation. The sample is counted in a plexiglass container placed between two 8- by 4-inch NaI(Tl) crystals. The data are recorded in a multichannel analyzer and processed by a computer code using standard spectra taken on the same counting apparatus.

The vegetation samples taken from nine locations throughout the Livermore Valley are dried in ambient air for about 1 week. Then 150–200 grams of dry material are pressed into a pellet 5 inches in diameter by ½ to 1 inch thick. The pellet is then counted on the system described for milk. The data are also processed by computer.

Samples of environmental water, milk, and vegetation are analyzed for tritium. One milliliter of water or distillate is counted in a liquid scintillation counter having a 32 percent efficiency for HTO.

The soil samples are prepared by digesting 1 gram of material in a mixture of perchloric and nitric acids and evaporating to near dryness. The resulting residue is redissolved in a hydrogen peroxide and nitric acid mixture and subsequently plated onto stainless-steel plates. The water and sewage samples, of ½ liter volume, are treated similarly.

2. Special measurements:

Soil samples are subjected to either a complete acid dissolution or the HASL[2] leach technique for plutonium-239 and strontium-90 determinations. A minimum of 100 grams of soil is used for these analyses. The cesium-137 activities in soil samples are determined directly by placing approximately 300 grams of soil on GeLi detector equipped with background and Compton suppression systems.

The water samples that are analyzed for tritium activity are pretreated by vacuum distillation under an argon atmosphere, enriched 75 fold electrolytically and counted by a liquid scintillation detector.

The air and water samples are subjected to gross alpha and gross beta measurements. Even though the analyses of these samples indicate that iodine-129, actinium-227, radium-226, and radium-228 are not present, the average gross alpha activities still may not exceed the AEC standards for air and water for controlled and uncontrolled areas. Since the alpha emitters have already been accounted for in the gross alpha measurements, the average gross beta radioactivity of the samples may not exceed the AEC standards for air and water for controlled and uncontrolled areas. The AEC Manual Chapter 0524 also states that the average tritium activities in offsite water samples may not exceed 3 μCi/liter.

The average monthly alpha and beta radioactivity of sewage effluent samples should not exceed the guide value of 400 pCi/liter (AEC standard for

[2] HASL–Health and Safety Laboratory, U.S. Atomic Energy Commission.

water in a controlled area), and the tritium activity should not exceed 10^8 pCi/liter.

The annual external radiation dose to the whole body of workers in controlled areas may not exceed 5 rem; while that to individuals in uncontrolled areas may not exceed 170 mrem.

Air and soil samples are collected to ascertain that control efforts are restricting the release of radioactivity from the laboratory to levels that do not exceed the permissible levels for the neighborhood around an atomic energy facility. The water samples are collected to ascertain that radioactivity is not entering the underground water supply which provides most of the domestic water for the cities of Livermore and Pleasanton, the sole supply for ranches in the Livermore and Amador Valleys. Samples were taken at the Livermore Sewage Disposal Plant to establish that liquid effluent from the laboratory is not contributing significant radioactivity to the dried sludge or liquid effluent produced. The dried sludge is used as an agricultural soil conditioner. Some liquid effluent released to the Los Positas Arroyo is used for irrigation of the golf course turf and some agricultural land.

Air particulate samples were collected continuously at 15 sampling stations. The samplers situated on the laboratory perimeter utilize 80 square inch Whatman filters at an average flow rate of 25 cfm. The remaining samplers utilize 36 square inch HV-70 filters at an average flow of 4 cfm. Filters are changed weekly during the winter months and biweekly during the summer to avoid excessive mass loading. The results are shown in table 6.

At no time was any detectable alpha radioactivity observed on the air filters. The average airborne beta radioactivity was 0.1 pCi/m^3; about the same as that reported during the previous 6 months.

Soil samples were collected quarterly. Only the top 1 cm layer was collected at the 19 sampling locations in the Livermore Valley. The alpha radioactivity of each sample was less than the

Table 6. Airborne particulate radioactivity around LRL Livermore site July–December 1970

Sampling location	Number of samples	Alpha radioactivity[a] (fCi/m³)		Beta radioactivity[a] (fCi/m³)	
		Average	Maximum	Average	Maximum
1	46	≤0.4	≤0.4	107	210
2	49	≤ .4	≤ .4	98	230
3	26	≤1.0	≤1.0	72	144
4	24	≤1.0	≤1.0	85	204
5	25	≦1.0	≤1.0	87	191
6[b]	NS				
7	26	≤1.0	≤1.0	64	142
8	26	≤1.0	≤1.0	69	205
9	22	≤1.0	≤1.0	119	330
10	23	≤1.0	≤1.0	66	130
11	25	≤1.0	≤1.0	53	99
12	49	≤ .4	≤ .4	121	248
13	49	≤ .4	≤ .4	109	264
14	49	≤ .4	≤ .4	152	420
15	48	≤ .4	≤ .4	130	319
16	26	≤1.0	≤1.0	93	213

[a] Approximate limits of sensitivity: Alpha—locations 1, 2, 12–15—4 fCi/m³.
locations 3–11, 16—1.0 fCi/m³.
Beta—locations 1, 2, 12–15—4 fCi/m³.
locations 3–11, 16—10 fCi/m³.
Estimated counting error limits: alpha radioactivity ±30 percent; beta radioactivity ±20 percent.
[b] Sampler is being relocated.

Table 7. Radioactivity in soil samples, July–December 1970

Location	Depth (inches)	Average concentration (dpm/g)		
		Plutonium-239	Cesium-137	Strontium-90
Onsite[a]	0–2	$2.4\times10^{-2}\pm 5\%$	0.86 ± 6%	—
	2–4	$1.8\times10^{-2}\pm 6\%$	.69 ±13%	—
	4–6	$2.3\times10^{-3}\pm 13\%$	.20 ±25%	—
Offsite	0–2	$2.8\times10^{-3}\pm 16\%$	.38 ±21%	—
	2–4	$1.8\times10^{-2}\pm 26\%$	.44 ±25%	—
	4–6	$3.7\times10^{-4}\pm 20\%$	.15 ±53%	—
Offsite	0–2	$3.8\times10^{-3}\pm 19\%$	.22 ±36%	—
Offsite	0–6	$9.0\times10^{-4}\pm 9\%$	≤ .03	0.099 ±11%
	0–6	$1.2\times10^{-3}\pm 10\%$	≤ .03	.061 ±18%
	0–6	$3.9\times10^{-3}\pm 5\%$	.081 ±22%	.077 ± 7%
	0–6	$2.8\times10^{-3}\pm 6\%$	.063 ±27%	.082 ± 7%
	0–6	$2.5\times10^{-3}\pm 6\%$	.050 ±32%	.055 ±14%
	0–6	$8.6\times10^{-4}\pm 11\%$	≤ .03	.045 ±14%

[a] Adjacent to area used for solar evaporation of radioactive liquids.

limit of detection (1.5 pCi/g). The beta radioactivities fluctuated from 4.7 pCi/g to 15 pCi/g with an average of 7.5 pCi/g. The limit of detection is 3.5 pCi/g. The radioactivity detected is within the normal range for soil in the Livermore Valley, as determined from surveillance during previous years.

Additional soil samples were collected onsite and offsite within a 4-square-mile area immediately east of the Livermore Laboratory. These samples were subjected to plutonium-239, strontium-90 or cesium-137 determinations. The first seven samples shown in table 7 were analyzed by complete acid dissolution of the soil while the remaining samples were subjected to the HASL acid leach technique for plutonium-239 determination. Each sample was collected over a 10 square-inch surface area.

Domestic water samples were collected monthly from 10 nearby sources. No sample showed an alpha radioactivity above the limit of detection (1.2 pCi/liter). The gross beta radioactivities are shown in table 8.

The limit of detection for gross beta radioactivity is 2.5 pCi/liter. The tritium activity of the water samples fluctuated from below the limit of detection (3.5 nCi/liter) to 5.4 nCi/liter. The average tritium activity was ≤3.6 nCi/liter. Several water samples were collected weekly

Table 8. Beta radioactivity in domestic water supplies LRL Livermore site, July–December 1970

Dosimeter location	Beta radioactivity[a] (pCi/liter)	
	Maximum	Average
1	3.4	3.4
3	7.7	2.8
4	5.9	≤2.5
5	4.1	≤2.5
6	3.2	≤2.5
7	14	4.1
8	5.5	≤2.5
10	4.5	≤2.5
Building 253	≤2.5	≤2.5
Building 419	≤2.5	≤2.5

[a] The limit of detection for gross beta radioactivity is 2.5 pCi/liter.

during the months of July, August, and September from the city of Livermore and Site 300 wells, as well as other miscellaneous sources of drinking

Table 9. Tritium radioactivity in water LRL Livermore, July–December 1970

Livermore		Site 300	
Well number	Tritium (pCi/liter)	Well number	Tritium (pCi/liter)
8	19±19	1	14±14
9	22±10	3	≤14
10	25±18	4	≤20
13	75±15	5	23±14
14	72±13	6	50±18
15	17±14	7	≤13
17	39±12	8	≤20
19	74±15		
20	80±17		

water in the immediate vicinity of the laboratory. The weekly samples collected at a particular source were composited into a single sample from that source. Tritium results are shown in table 9.

Samples were collected daily at the laboratory sewer discharge. A portion of each of these daily samples were composited during each month. These monthly composite samples for the radionuclides analyzed are shown in table 10.

When the sewage from the laboratory and city of Livermore enters the sewage disposal plant, the liquid and sludge are separated, and the sludge passes into one of two digesters where it is broken down by bacterial action. Methane gas is evolved and burned and the remaining sludge is released to large, sludge lagoons and/or drying beds and

Table 10. Radioactivity in laboratory sewer effluent samples July–December 1970

Month (1970)	Alpha radioactivity (pCi/liter)		Beta radioactivity (pCi/liter)		Tritium (nCi/liter)		Strontium-90 (pCi/liter)	Radium-226 (fCi/liter)	Radium-228 (pCi/liter)	Mixed iodine (pCi/liter)
	Maximum	Average	Maximum	Average	Maximum	Average				
July	24	8.0	1,040	157	500	22	5.1	34	0.36	0.34
August	18	4.0	1,724	387	68	6	22	120	1.1	3.8
September	9	≤1.2	213	102	21	4.6	4.9	45	.37	60
October	1.2	≤1.2	210	75	14	4.7	4.0	18	.35	25
November	19	4.0	285	105	24	7.5	7.4	150	1.7	11
December	19	7.0	373	77	82	25	4.5	35	.39	9.1

Table 11. Radioactivity in samples from the Livermore sewage disposal plant, July–December 1970

Radioactivity, units of measurement and location of sampling	1970					
	July	August	September	October	November	December
Gross alpha (pCi/liter)						
Digesters:						
Maximum	452	455	321	508	220	338
Average	306	290	254	276	144	214
Aeration tank:						
Maximum	201	120	150	33	41	86
Average	163	107	64	24	27	42
Effluent:						
Maximum	≤1.2	≤1.2	≤1.2	≤1.2	≤1.2	≤1.2
Average	≤1.2	≤1.2	≤1.2	≤1.2	≤1.2	≤1.2
Gross beta (pCi/liter)						
Digesters:						
Maximum	2,000	5,500	1,200	1,300	6,200	1,800
Average	1,300	1,700	830	800	500	750
Aeration tank:						
Maximum	325	329	417	84	70	103
Average	281	213	206	67	55	58
Effluent:						
Maximum	45	130	45	28	25	32
Average	28	46	19	13	8	13
Tritium (nCi/liter)						
Aeration tank:						
Maximum	4.0	5.0	≤3.5	≤3.5	≤3.5	33
Average	≤3.6	≤3.9	≤3.5	≤3.5	≤3.5	≤9.7
Effluent:						
Maximum	29	7.0	≤3.5	4.0	≤3.5	12
Average	≤6.6	≤3.9	≤3.5	≤3.5	≤3.5	≤5.8
Strontium-90 (pCi/liter)						
Effluent:						
Average	NA	NA	≤ .4	≤ .4	1.0	.5

NA, no analysis.

retained for subsequent possible use as a soil conditioner. The purified water is used on the Livermore golf course and agricultural irrigation on a nearby farm; the excess is discharged into the Los Positas Arroyo. Samples were collected weekly from the aeration tanks, each digester, and the effluent to the arroyo. The results of these analyses are given in table 11. The limits of detection are 1.2 pCi/liter for the gross alpha radioactivity, 2.5 pCi/liter for the gross beta measurements, 3.5 nCi/liter for the tritium determinations, and 0.4 pCi/liter for strontium-90.

Samples were collected monthly from each of two dairies in the Livermore Laboratory and Site 300 areas. The average radioactivity in the milk samples is shown in table 12.

Table 12. Radionuclides in milk, LRL July–December 1970

Location of dairy	Concentration (pCi/liter)		
	Cesium-137	Cerium-144	Tritium
3 miles north of Site 300	≤1.9	≤3.5	≤3,500
6 miles west of Livermore Laboratory	2.7	≤5.5	≤3,500

Samples of vegetation were collected during July and October of 1970; the radioactivity per unit dry weight of unwashed grass samples and eucalyptus leaf samples is shown in table 13. A portion of each sample was freeze dried and the water collected was analyzed for HTO.

Continuous air sampling is maintained on the stack effluent from operations that are potential sources of radioactive material releases. These facilities include hot cells, transuranic chemistry, plutonium metallurgy, and the reactor building. The effluent from these buildings passes through high-efficiency filters prior to being exhausted to the atmosphere. The offsite concentrations were not exceeded as a result of the stack effluent from these facilities.

Table 13. Radioactivity in vegetation, LRL Livermore site, July–December 1970

Nuclide	Concentration (pCi/g dry weight)				Limit of detection
	Grass		Eucalyptus leaves		
	Average	Maximum	Average	Maximum	
Beryllium-7	1.1	3.1	0.58	1.2	0.01
Manganese-54	.18	1.1	≤ .01	≤ .01	.01
Zirconium-95	.49	1.6	.16	.37	.01
Ruthenium-103	≤ .02	.04	≤ .02	.04	.02
Ruthenium-106	.56	2.7	.31	.96	.02
Antimony-125	≤ .02	.03	≤ .02	.09	.02
Cesium-137	.04	.12	.06	.13	.01
Barium-140	.52	1.9	.07	.11	.01
Cerium-144	1.5	4.1	.64	1.3	.02
Tritium (nCi/liter)	14	22	5.8	17	3.5

The measurement of low-level gamma radiation was accomplished with thermoluminescent dosimeters at nine points on the site perimeter and at three offsite locations. The dose received by each detector during the 6-month period is shown in table 14. The limit of detection of these dosimeters is 10 mR.

Accidential release of tritium gas

Continuous air sampling is also maintained on the stack effluent from the gaseous chemistry Building. On August 6, 1970, an accidental release of tritium gas occurred from this facility. During the following 4 days, an extensive environmental sampling program was conducted to evaluate the effects of the release on the environment. The vast amount of environmental data that was obtained is too voluminous to report here and will be reported shortly in a separate publication.

The limits of detection for tritium is 5 nCi/liter of water and 40 pCi/m^3 of air. The data indicate that no significant exposure to onsite or offsite individuals resulted from the release. Detectable activities were observed in some vegetation and atmospheric water vapor samples. However, even the highest measured concentrations, as well as the

Table 14. External gamma radiation measurements LRL Livermore site, July–December 1970

Dosimeter locations[a]	Exposure[b] (mR)
1	17
2	≤ 10
3	≤ 10
4	≤ 10
5	70
6	≤ 10
7	17
8	14
9	15
10	≤ 10
11	≤ 10
12	≤ 10

[a] Locations 1–9, controlled area perimeter. Locations 10, 11, and 12, uncontrolled area.
[b] Minus backgroud.

calculated maximum credible ground level air concentration, that could have resulted during cloud passage were well below the offsite maximum permissible concentration for continuous exposure.

Site 300

The high explosive test area at Site 300 (figure 3) covers 10-square miles in the hills, in a very sparsely populated ranching area about 15 miles southeast of Livermore and 8 miles southwest of Tracy. The chief area of concern for Site 300 is Tracy with a population of about 15,000. Air and water samples are taken to determine whether operations at Site 300 are changing the normal radioactivity levels in the vicinity. The 11 air samplers at Site 300 are operated on a continuous basis at about 25 cfm and the filter papers are

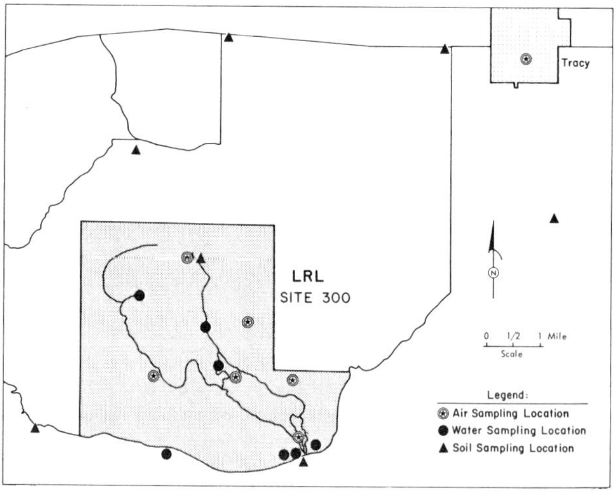

Figure 3. Sampling location at Site 300, LRL

Table 15. Airborne particulate radioactivity, LRL Site 300, July–December 1970

Sampling location	Number of samples	Alpha radioactivity[a] (fCi/m^3)		Beta radioactivity[a] (fCi/m^3)	
		Average	Maximum	Average	Maximum
1[b]	34	<0.2	0.2	161	378
2	31	< .2	1.0	210	418
3	35	< .2	1.0	181	472
4	30	< .2	< .2	192	386
5	34	< .2	6.6	136	304
6[b]	10	< .2	< .2	132	212
7	34	<1.0	<1.0	148	332
8	33	< .2	< .2	164	430
9	34	< .2	1.1	233	521
10	23	< .2	1.0	180	399
11	33	< .2	< .2	147	372
Summary	331	<0.2	6.6	171	521

[a] Limit of detection: alpha radioactivity, location 1–6, 8–11, 0.2 fCi/m^3; location 7, 1.0 fCi/m^3; beta radioactivity, location 1–6, 8–11, 2 fCi/m^3; location 7, 10 fCi/m^3.
Variations in limits of detection results from differing flow rates. Estimated counting error limits, alpha radioactivity ±30 percent; beta radioactivity ±20 percent.
[b] Sampler inoperative for several weeks.

changed weekly. Most of these air samplers are located within the boundaries of the test site due to unavailability of electric power facilities offsite. Table 15 gives the airborne particulate radioactivity for July–December 1970. Water samples are collected monthly from onsite wells. Streams in the area are sampled when they contain water (usually during the winter). No sample showed an alpha radioactivity concentration above the limit of detection (1.2 pCi/liter). The average beta concentrations are given in table 16. The tritium activities of the water samples fluctuated from below the limits of detection (3.5×10^3 pCi/liter) to a high of 8.6×10^3 pCi/liter. The average tritium activity was $\leq 3.8 \times 10^3$ pCi/liter.

Table 16. Beta radioactivity in water samples, LRL Site 300, July–December 1970

Sampling location	Beta radioactivity (pCi/liter)	
	Maximum	Average
1	3.6	2.6
2	6.8	4.7
3	9.5	6.9
4	8.6	4.9
5	10.9	5.7
6	6.4	3.4
7	6.8	5.2
11	10.9	7.9
14	7.7	6.2

Soil samples were collected quarterly at nine offsite locations. The alpha radioactivity was below the limit of detection (1.5 pCi/g). The beta radioactivity ranged from the limit of detection (3.5 pCi/g) to 15 pCi/g, with an average activity of 5.9 pCi/g.

None of the average concentrations of air, water, and soil samples exceeded the concentration guide levels of environmental radioactivity, nor did any of the results increase significantly compared to the previous 6-month reporting period. Therefore, it appears that radiological control measures at Site 300 were adequate and that no significant radioactivity was released to the environment near the site during this period.

Preoperational Radiological Surveillance of the Florida Power Corporation's Crystal River Power Reactor Site

Chester L. Nayfield

Wallace Johnson

Benjamin P. Prewitt

Preoperational radiological surveillance has been conducted around the Crystal River site of the Florida Power Corporation since May 1969 by the Florida Division of Health. Present activities at this site include a 387 MWe coal-fired generating unit and a 510 MWe coal-fired generating unit. Construction of unit three is underway. This unit is to be an 855 MWe nuclear power generating plant. Locations of Division of Health sampling sites are shown in figure 1. During 1969–1970, samples of environmental media shown in table 1 were collected.

Table 1. Number of environmental samples collected Crystal River

Vector	Number of samples		
	1969	1970	Total
Soil	79	68	147
Silt	16	14	30
Vegetation	79	126	205
Surface water	58	49	107
Drinking water	24	35	59
Marine biota	4	20	24
Milk	NS	2	2
Citrus	NS	5	5

NS, no sample.

Analyses of samples were conducted in the Radiological Laboratory of the Division of Health located in Orlando. Analytical instruments used included a 1,024 channel analyzer using a 4- by 4-inch NaI detector, a low-beta internal proportional counter, a beta internal proportional counter with automatic sample changer, and a Tri-Carb liquid scintillation spectrometer. The minimum detectable activities used in this laboratory for the various sample types are shown in table 2.

Table 2. Minimum detectable activities, Crystal River

Sample type	Alpha radioactivity	Beta radioactivity
Air	—	3.2 pCi/sample
Milk	—	1.2 pCi/liter
Soil and silt	3.4 pCi/g ash	1.1 pCi/g ash
Vegetation and biota	3.4 pCi/g ash	.78 pCi/g ash
Water	3.4 pCi/liter	.78 pCi/liter

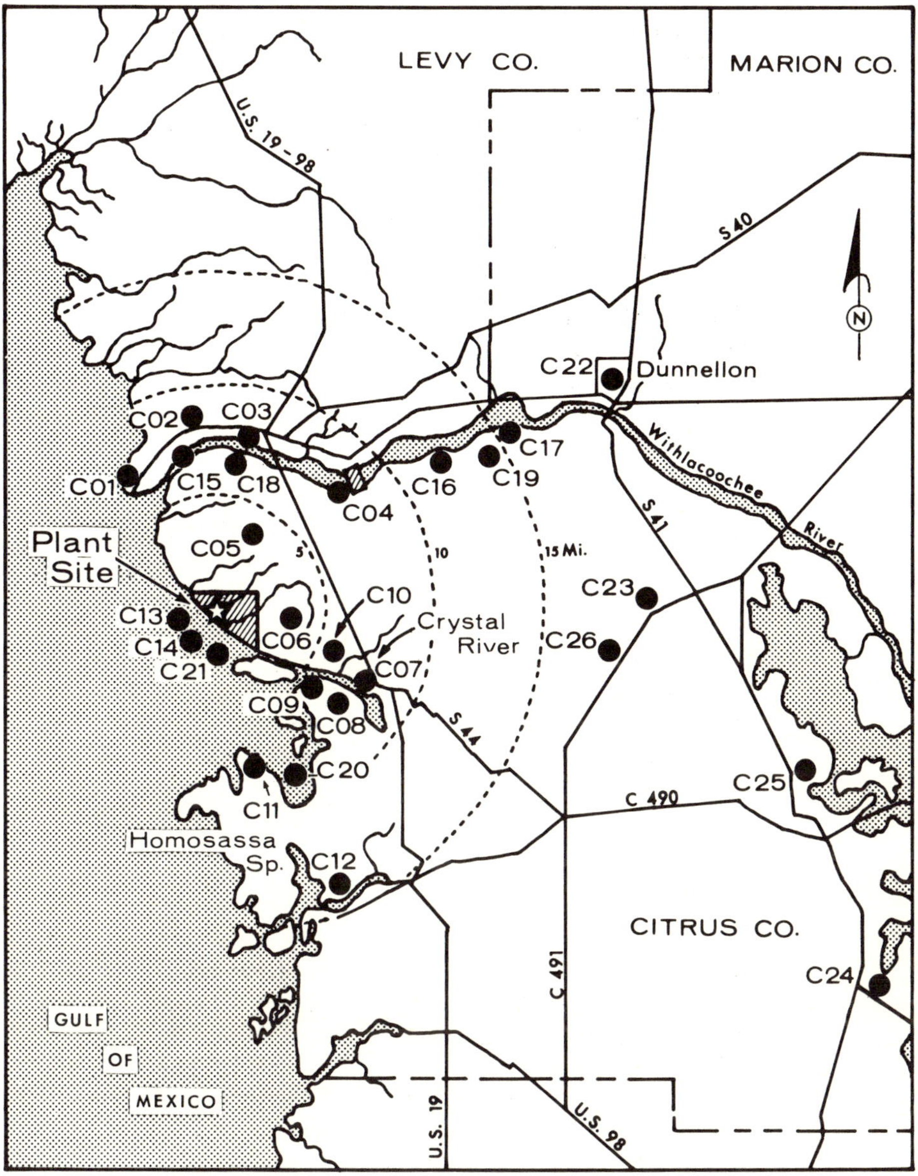

Figure 1. Radiological sampling sites, Crystal River

For spectroscopy, a 9 by 9 matrix solution has been used in analyzing the samples. For soil and silt samples, the matrix was expanded to 11 by 11 to correct the spectum for the radium-thorium contribution. Minimum detectable activities for the various nuclides included in the 9 by 9 matrix are shown in table 3.

The selection of sampling media around the Crystal River site was dictated by the nature of the region. Since all of the Division of Health's surveillance is human-health oriented, every effort was made to confine sampling media as directly as possible to the human food-chain. This was difficult for the Crystal River area. The closest dairy is 15 miles from the plant site. This dairy is sampled although its contribution to the milkshed is of minimal significance. Row crops are not grown in the area other than on a home garden basis. A citrus grove, located about 75 miles from the plant site, is sampled. The cabbage palm (*Sabal palmetto*) and the saw palmetto (*Serenoa serrulata*) were selected as indicators in the terrestrial food chain. Cabbage palm hearts are a delicacy in Florida and are extensively eaten by some population groups. The saw palmetto and its berries constitute an element in the diet of deer, which have been found to have

Table 3. Minimum detectable activities for 9 by 9 matrix

Nuclide	Minimum detectable activity[a] (1 liter geometry)
Cerium-144	98.6 pCi
Iodine-131	18.5 pCi
Ruthenium-106	78.5 pCi
Cesium-137	17.4 pCi
Zirconium-niobium-95	14.6 pCi
Manganese-54	15.2 pCi
Zinc-65	32.0 pCi
Potassium-40	19.4 pCi
Barium-lanthanum-140	18.8 pCi

[a] A practical reporting level of the minimum detectable activity plus or minus 2 sigma is used for reporting the sampling data.

elevated body burdens of cesium-137 (*1*). Deer meat is in the direct human food chain.

Data for 1969 and 1970 for soil samples and vegetation samples are presented in tables 4 and 5.

An analysis of variance was made of these data for 1970 and the values obtained are shown in table 6. From these analyses it was concluded that the variance between sampling sites of soil samples is significant only for cesium-137. Significant differences are apparent for all nuclides in vegetation. Only gross beta radioactivity levels are not significantly different in vegetation.

Analysis of variance of vegetation samples by subgroups of sampling sites is of interest (table 7). For cesium-137 in vegetation there is no significant difference in sites 5–11 and the variation of significance is in group 1–4.

A comparison of sample means by year is of interest. See tables 8 and 9.

Utilizing the sign test statistic (*2*) and a one-sided test at a level of 0.05, it can be demonstrated that a statistically significant decrease took place in soil between 1969 and 1970 for all radionuclides, except cerium-144. Utilizing the same statistic, vegetation samples show a decrease in radionuclides except for cesium-137 and the naturally occurring potassium-40. Based on the physical half-life of these radionuclides (zirconium-95, 65.5 days; ruthenium-106, 368 days; cerium-144, 284 days; cesium-137, 30.0 years), this behavior is expected with the exception of the retention of cerium-144 in soil samples.

A comparison of these data with those obtained in the surveillance of the Cape Kennedy offsite environs is interesting. Unfortunately, the Cape Kennedy program was inactivated in early 1970 so that data are available for 1969 only. Examination of table 10 indicates one very striking difference between soil samples collected in the Crystal River program and those collected in the Cape Kennedy offsite programs. Potassium-40

Table 4. Concentration of radionuclides in soil samples, Crystal River, 1969–1970

Sampling site and year	Radionuclide concentration (pCi/kg wet weight)									Radioactivity (pCi/g ash)	
	^{144}Ce	^{131}I	^{106}Ru	^{137}Cs	^{95}Zr	^{54}Mn	^{65}Zn	^{40}K	^{140}Ba	Gross beta	Gross alpha
C01:											
1969 Range	540–1,100	ND	200–520	ND–130	ND–300	ND	ND	ND	ND	<10	<7
Mean	562		320	104	167						
1970 Range	400– 730	ND	ND	ND– 80	ND– 90	ND–30	ND	ND	ND	<10	<7
Mean	499			40	52	[a]30					
C02:											
1969 Range	380–1,700	ND	ND–1,500	70–580	40–1,200	ND	ND	ND	ND	<10	<7
Mean	871		683	411	411						
1970 Range	290– 770	ND	ND–440	ND–530	ND–200	ND	ND	ND	ND	<10	<7
Mean	534		330	364	114						
C03:											
1969 Range	430–1,000	ND	ND–570	100–830	40–330	ND–30	ND	ND	ND	<10	<7
Mean	747		366	456	168	[a]30					
1970 Range	ND–650	ND	ND–200	60–440	ND–110	ND	ND	ND	ND	<10	<7
Mean	500		[a]200	211	53						
C04:											
1969 Range	440–1,300	ND	ND–910	270–620	ND–510	ND	ND	ND	ND	<10	<7
Mean	715		318	451	170						
1970 Range	280– 780	ND	ND–390	120–680	30–150	ND	ND	ND	ND	<10	<7
Mean	558		287	427	82						

C05:											
1969 Range	750–1,500	ND	ND–660	100–1,300	60–320	ND	ND	ND	ND	<10	<7
Mean	960		308	616	141						
1970 Range	440–1,400	ND	ND–370	340–1,000	ND–180	ND	ND	ND	ND	<10	<7
Mean	751		300	587	89						
C06:											
1969 Range	530–2,000	ND–100	ND–1,000	ND–1,900	ND–490	ND–60	ND	ND	ND	<10	<7
Mean	1,204	[a]100	476	1,034	221	50					
1970 Range	340– 900	ND–80	ND–740	ND–1,500	ND–100	ND	ND	ND	ND	<10	<7
Mean	651	60	340	783	80						
C08:											
1969 Range	360–1,900	ND	ND–1,700	ND–60	ND–1,200	ND	ND	ND	ND	<10	<7
Mean	875		633	56	431						
1970 Range	390–1,800	ND–100	ND–760	ND–2,100	ND–260	ND–50	ND	ND–550	ND	<10–62	<7–18
Mean	1,221	85	580	1,415	200	43		482		38	15
C09:											
1969 Range	490–1,400	ND	ND–1,300	ND	40–1,100			ND		<10	<7
Mean	716		553		291						
1970 Range	ND–720	ND	ND–380	ND–390	ND–150			ND		<10	<7
Mean	553		[a]380	[a]390	64						
C11:											
1969 Range	580–1,100	ND	ND–1,100	ND–420	50– 900	ND	ND	ND	ND	<10	<7
Mean	725		340	354	165						
1970 Range	540– 770	ND	ND–320	120–450	40–110	ND	ND	ND	ND	<10	<7
Mean	636		247	321	69						
C12:											
1969 Range	390–2,100	ND	ND–1,600	ND–580	40–1,200	ND–40	ND	ND	ND	<10	<7
Mean	921		441	372	294	[a]40					
1970 Range	ND–1,500	ND	ND–450	ND–1,000	ND–200	ND–50	ND	ND–540	ND	<10	<7
Mean	1,050		317	486	100	[a]50		520			

[a] Only one observation above detectable level.
ND, nondetectable.

Table 5. Concentration of radionuclides in vegetation samples, Crystal River, 1969–1970

Sampling site and year	Radionuclide concentration (pCi/kg wet weight)									Radioactivity (pCi/g ash)	
	^{144}Ce	^{131}I	^{106}Ru	^{137}Cs	^{95}Zr	^{54}Mn	^{65}Zn	^{40}K	^{140}Ba	Gross beta	Gross alpha
C01:											
1969 Range	ND–590	ND	40–1,460	90–210	290–2,100	ND	ND	3,500–7,600	ND	133–281	<7–16
Mean	510		789	159	951			5,825		202	10
1970 Range	ND–1,300	ND	ND–850	130–500	30–1,100	ND	ND	3,700–8,700	ND	73–301	<7–24
Mean	835		515	452	360			5,461		167	15
C02:											
1969 Range	ND	ND	ND–570	370–1,500	70–1,100	ND	ND	5,600–7,400	ND	210–294	<7–19
Mean			483	714	452			6,443		244	10
1970 Range	ND	ND	ND–710	220–900	70–580	ND	ND	4,800–6,700	ND	112–255	<7–30
Mean			430	564	322			5,867		180	17
C03:											
1969 Range	ND–1,400	ND	ND–1,700	180–760	90–2,600	ND	ND	2,400–5,100	ND	124–193	<7–17
Mean	[a]1,040		726	410	1,255			3,087		163	10
1970 Range	ND–540	ND	ND–1,000	ND–370	ND–960	ND	ND–120	3,200–6,600	ND	74–272	<7–26
Mean	[b]540		573	233	297		[b]120	5,427		154	14
C04:											
1969 Range	ND	ND	ND–110	1,100–4,400	160–940	ND	ND	2,500–6,200	ND	63–314	<7–19
Mean			[b]110	2,475	412			3,687		174	12
1970 Range	ND	ND	ND–630	340–3,500	130–740	ND	ND	1,200–5,600	ND	93–243	<7–38
Mean			465	1,735	413			3,238		164	17

C05:										
1969 Range	ND–400	ND	350–1,500	390–1,400	310–2,000	ND	ND	4,000–7,400	ND	68–264	<7–22
Mean	[b]400		612	829	949			5,237		205	13
1970 Range	ND	ND	360–680	110–1,300	190–740	ND	ND	1,300–7,300	ND	69–207	<7–30
Mean			490	526	437			3,584		135	15
C06:											
1969 Range	ND	ND	ND–1,300	180–6,300	170–2,100	ND	ND	1,600–5,800	ND	109–229	<7–15
Mean			568	3,069	847			3,285		197	9
1970 Range	ND–390	ND	ND–360	170–4,600	70–510	ND–60	ND	3,000–7,300	ND	90–287	9–42
Mean	373		330	333	296	[b]60		4,918		187	16
C08:											
1969 Range	ND–110	ND	270–980	ND–250	240–1,400	ND	ND	3,900–6,700	ND	106–202	<7–13
Mean	[b]110		662	128	625			5,213		150	9
1970 Range	ND–350	ND	ND–700	ND–160	140–680	ND	ND	4,000–6,800	ND	72–215	<7–32
Mean	[b]350		526	103	346			4,940		147	13
C09:											
1969 Range	ND	ND	320–1,600	ND–140	150–770	ND	ND	4,900–7,100	ND	141–241	<7–10
Mean			759	110	549			5,938		178	8
1970 Range	ND	ND	ND–940	ND–160	ND–770	ND	ND	ND-7,000	ND	97–243	<7–26
Mean			507	102	369			5,236		164	13
C11:											
1969 Range	ND–730	ND	520–1,800	ND–160	280–3,000	ND	ND	3,600–6,600	ND	183–277	<7–21
Mean	526		1,077	131	1,331			5,225		233	13
1970 Range	ND–1,000	ND	360–860	80–280	230–1,300	ND	ND	3,400–6,900	ND	125–264	<7–26
Mean	750		619	152	693			5.058		179	17
C12:											
1969 Range	ND–430	ND	ND–870	100–330	180–1,800	ND	ND	2,600–7,500	ND	138–236	<7–14
Mean	[b]430		638	180	761			5,368		168	9
1970 Range	ND	ND	ND–510	ND–580	90–420	ND	ND	2,800–7,500	ND	104–234	<7–27
Mean			395	208	240			5,963		160	16

[a] Mean of two observations above detectable level.
[b] Only one observation above detectable level.
ND, nondetectable.

has been essentially nondetectable in the Crystal River samples while it was detectable in nearly all of the Cape Kennedy samples. Utilizing the the Mann-Whitney "U" test (*3*), there is no significant difference between the cesium-137 levels in soil samples from the two programs at a 95-percent confidence level.

There is a significant difference in the means of cesium-137 levels between vegetation samples from Crystal River and Cape Kennedy. No significant difference is found between potassium-40 levels or between gross beta radioactivity levels. It is noted, however, that for cesium-137 there is no significant difference between the Cape Kennedy Zone A samples and the Crystal River samples. Zone A of the Cape Kennedy sampling plan is within a 10-mile radius of the launch site (figure 2) and corresponds in location with a majority of Crystal River sites (figure 1). The higher concentrations of cesium-137 occurred in Zone B (20 miles) and Zone C (30 miles) at Cape Kennedy. We conclude, therefore, that the elevated levels of cesium-137 demonstrated in palmetto at sites 4, 5, and 6 at Crystal River in 1969 and at sites 4 and 5 in 1970, are representative of more inland locations such as those of Zone B and Zone C at Cape Kennedy.

Table 6. Results of the analysis of variance

Nuclide	Soil		
	Calculated F	95 percent F value	Conclusion
Cerium-144	0.0	2.25	Not significant
Cesium-137	7.5	2.25	Significant
Zirconium-95	1.3	2.25	Not significant
	Vegetation		
Cesium-137	21.1	2.01< F<2.09	Significant
Zirconium-95	2.4	2.01< F<2.09	Significant
Potassium-40	6.0	2.01< F<2.09	Significant
Gross beta	.27	2.01< F<2.09	Not significant

Table 7. Analysis of variance of vegetation samples by subgroups of sampling sites, Crystal River

Radionuclide	Calculated F	95 percent F value	Conclusion
Cesium-137:			
Sites 1–4	16.3	2.76<F<2.84	Significant
Sites 5–11	1.40	2.53<F<2.61	Not significant
Zirconium-95:			
Sites 1–4	.14	2.76<F<2.84	Not significant
Sites 5–11	3.97	2.53<F<2.61	Significant
Potassium-40:			
Sites 1–4	12.8	2.76<F<2.84	Significant
Sites 5–11	2.73	2.53<F<2.61	Significant
Gross beta:			
Sites 1–4	2.39	2.76<F<2.84	Not significant
Sites 5–11	.57	2.53<F<2.61	Not significant

Table 8. Soil sample means by radionuclide and by year, Crystal River, 1969–1970

Sampling site and year	Sample mean ($\overline{X}$)[a] (pCi/kg wet weight)			
	Cerium-144	Ruthenium-106	Cesium-137	Zirconium-95
C01:				
1969	562	320	104	167
1970	499	ND	40	52
C02:				
1969	871	683	411	411
1970	534	330	364	114
C03:				
1969	747	366	456	168
1970	500	200	211	53
C04:				
1969	715	318	451	170
1970	558	287	427	82
C05:				
1969	960	308	616	141
1970	751	300	587	89
C06:				
1969	1,204	476	1,034	221
1970	651	340	783	80
C08:				
1969	875	633	56	431
1970	1,221	580	1,415	200
C09:				
1969	716	553	ND	291
1970	553	380	390	64
C11:				
1969	725	340	354	165
1970	636	247	321	69
C12:				
1969	921	441	372	294
1970	1,050	317	486	100
$\overline{\overline{X}}$:				
1969	830	444	428	246
1970	695	331	502	83

[a] Nondetectable sample results not included in mean.
ND, all sample results were nondetectable.

Table 9. Vegetation sample means by radioactivity and by year, Crystal River, 1969–1970

Sampling site and year	Sample mean ($\bar{\bar{X}}$) (pCi/kg wet weight)				Sample mean ($\bar{\bar{X}}$)	
	Ruthenium-106	Cesium-137	Zirconium-95	Potassium-40	Gross beta (pCi/g ash)	Gross beta[a] (pCi/kg wet weight)
C01:						
1969	789	159	951	5,825	202	6,052
1970	515	452	360	5,461	167	5,305
C02:						
1969	483	714	452	6,463	244	6,731
1970	430	564	322	5,867	180	5,484
C03:						
1969	726	410	1,255	3,087	163	4,620
1970	573	233	297	5,427	154	4,915
C04:						
1969	110	2,475	412	3,687	174	4,399
1970	465	1,735	413	3,238	164	3,841
C05:						
1969	612	829	949	5,237	205	5,754
1970	490	526	437	3,584	135	3,968
C06:						
1969	568	3,069	847	3,285	197	5,124
1970	330	333	296	4,918	187	4,840
C08:						
1969	662	128	625	5,213	150	5,115
1970	526	103	346	4,940	147	5,583
C09:						
1969	759	110	846	5,938	178	6,048
1970	507	102	369	5,236	164	5,255
C11:						
1969	1,077	131	1,331	5,225	233	5,766
1970	619	152	693	5,058	179	5,122
C12:						
1969	638	180	761	5,368	168	5,603
1970	395	208	240	5,963	160	4,957
$\bar{\bar{X}}$:						
1969	642	819	713	4,933	191	5,521
1970	485	441	377	4,969	164	4,927

[a] Not reported in table 5.

Table 10. Cape Kennedy offsite data compared to Crystal River data for 1969

Location	Soil (pCi/kg wet weight)					Vegetation				
						Gross beta (pCi/g ash)	(pCi/Kg wet weight)			
	^{144}Ce	^{106}Ru	^{137}Cs	^{95}Zr	^{40}K		^{106}Ru	^{137}Cs	^{95}Zr	^{40}K
Zone A composite	1,346	270	193	100	1,220	233	1,500	600	2,000	5,800
Zone B composite	779	315	244	87	746	195	990	2,770	805	3,075
Zone C composite	1,302	254	324	116	1,092	188	995	1,500	623	2,866
Cape Kennedy grand mean	1,142	279	254	101	1,019	208	1,161	1,623	1,143	3,910
Crystal River grand mean	830	444	428	246	[a]501	191	642	819	713	4,933
Division of Health Bureau of Sanitary Engineering sampling						178	890	1,080	773	4,666

[a] Two sampling locations only; all other samples were nondetectable.

The levels of radionuclides in vegetation taken at Crystal River in 1969 show a much closer correlation to samples taken in 1969 by the Bureau of Sanitary Engineering of the Division of Health in Seminole and Volusia Counties than to those from the Cape Kennedy area. The Seminole County and Volusia County samples were analyzed by the Orlando Radiological Laboratory as were the Crystal River samples. Utilizing the "Student's t" distribution at a 95-percent confidence level, there is no significant difference between the means of cesium-137 from the respective locations.

Levels of radioactivity in water samples are shown in table 11. Levels of activity in drinking water derived from wells have been nondetectable for 1969 and 1970 (table 12). Surface water samples are taken over a wide range of sources with varying salinities. Undissolved solid gross beta radioactivity levels have been nondetectable for all of the samples taken in 1969 and 1970. Levels of gross beta radioactivity in dissolved solids are dependent on the salinity of the water and have ranged from 5 to 495 pCi/liter. Potassium-40 is the only radionuclide which has been found in water in detectable concentrations by gamma analysis, and only at site C13 (intake canal) and site C14 (discharge canal). The normal level of potassium-40 in seawater of 35 parts per thousand salinity is reported to be 330 pCi/liter (*4*). There were no detectable concentrations of tritium (minimum detectable concentration = 200 pCi/liter) in any of the water samples.

Levels of radioactivity detected in various marine biota in 1970 are outlined in table 13. Levels of the various radionuclides from fish samples may be compared to those detected in fish samples taken from the Turkey Point surveillance program (table 14). It is noted that there is a good agreement between means of potassium-40 as would be expected. Testing the means of gross beta with the "Student t" distribution (*5*) at a 95-percent confidence level indi-

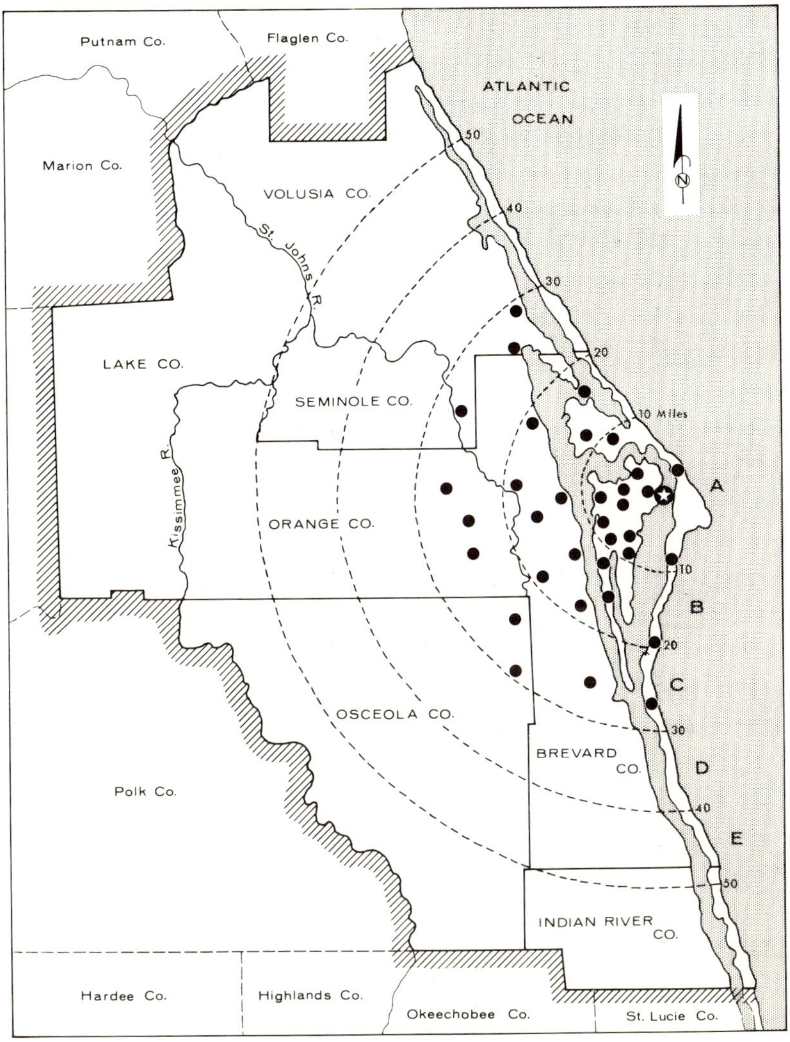

Figure 2. Cape Kennedy fixed offsite sampling areas

Table 11. Sample means of gross beta radioactivity in surface water samples, Crystal River, 1969–1970

Sampling location	1969			1970		
	Undissolved solids (pCi/liter)	Dissolved solids (pCi/liter)		Undissolved solids (pCi/liter)	Dissolved solids (pCi/liter)	
	Mean	Mean	Range	Mean	Mean	Range
C01:						
Mouth of Withlacoochee	<10	50	5–137	<10	48	19–130
C08:						
Tide Flat	<10	39	10–96	<10	23	10–51
C09:						
Crystal River	<10	104	56–162	<10	58	26–73
C11:						
Open Gulf	<10	143	99–207	<10	109	43–193
C12:						
Homasassa River	<10	25	8–101	<10	9	5–15
C13:						
Intake canal	<10	249	112–495	<10	152	123–214
Potassium-40		270	220–310		223	190–280
C14:						
Discharge canal	<10	201	127–430	<10	210	52–271
Potassium-40		232	ND–310		253	240–280
C16:						
Withlacoochee River	<10	<10		<10	<10	
C17:						
Withlacoochee River	<10	<10		<10	<10	

Table 12. Sample means of gross beta radioactivity in drinking water samples, Crystal River, 1969–1970

Sampling location	1969		1970	
	Undissolved solids (pCi/liter)	Dissolved solids (pCi/liter)	Undissolved solids (pCi/liter)	Dissolved solids (pCi/liter)
C07: Crystal River City well	<10	<10	<10	<10
C10: Indian waters well	<10	<10	<10	<10
C18: Yankeetown City well	<10	<10	<10	<10
C24: City of Inverness well			<10	<10
C22: City of Dunnellon well			<10	<10
C23: Holder deep well			<10	<10

cates a significant difference. The difference between means of cesium-137 is apparent. Other radionuclides were generally nondetectable in both programs.

A comparison of oyster samples collected at the Crystal River site in 1970 may be made with samples taken from Apalachicola Bay during the same period (table 15).

Conclusion

A number of conclusions may be drawn from the data presented relative to the preoperational distribution of radionuclides in the Crystal River area.

1. There has been a reduction in the levels of artificial radionuclides present in the environment except for the long half-life cesium-137.

Table 13. Levels of radionuclides detected in marine biota, Crystal River, 1970

Sampling site and vector	Radioactivity (pCi/kg wet weight)					
	^{144}Ce	^{106}Ru	^{137}Cs	^{95}Zr	^{40}K	Beta[a] (pCi/g ash)
Crab:						
C01	200	ND	210	100	2,000	<10
C21	640	ND	90	ND	2,000	14
	220	ND	ND	40	1,500	<10
C20	600	ND	110	30	2,800	21
C13	310	ND	90	80	1,700	16
	ND	ND	ND	ND	1,400	20
Mean	394	ND	125	62	1,900	15
Oysters:						
C14	ND	ND	ND	30	2,300	11
	ND	ND	80	ND	2,500	57
	ND	360	ND	ND	1,900	11
C13	ND	ND	ND	ND	ND	52
C11	ND	ND	ND	30	1,200	87
C08	ND	ND	ND	ND	ND	43
Mean	ND	360	80	30	1,975	43
Fish:						
C21	ND	ND	ND	ND	1,200	24
Redfish:						
C20	ND	ND	ND	ND	2,300	41
C14	ND	ND	ND	ND	2,700	52
Mullet:						
C01	ND	ND	ND	ND	2,400	70
C11	740	ND	130	ND	3,700	59
	ND	ND	ND	ND	2,700	26
	ND	ND	ND	ND	2,000	67
Mean	740	ND	130	ND	2,428	48

[a] Beta, <10 pCi/g ash averaged as 10 pCi/g ash.
ND, nondetectable, not included in average.

Table 14. Comparison of mean levels of radionuclides in fish samples, Crystal River—Turkey Point, 1970

Radioactivity	Crystal River mean	Turkey Point mean
Potassium-40 (pCi/kg wet weight)	2,428	2,511
Gross beta (pCi/g ash)	48	36
Cesium-137 (pCi/kg wet weight)	[a]130	68

[a] Only one detectable observation.

Table 15. Comparison of oyster samples, 1970

Radioactivity	Crystal River mean	Apalachicola Bay mean
Gross beta (pCi/g ash)	43 (6 samples)	49 (4 samples)
Potassium-40 (pCi/kg wet weight)	1,975	1,946

There is no significant difference between the means for gross beta radioactivity or between the means for potassium-40.

2. Levels of radionuclides in soil are not significantly different from control locations (Cape Kennedy offsite sampling sites) except for levels of potassium-40 which appear to be lower than those of the control locations.

3. Levels of radionuclides in vegetation are not significantly different from levels of comparable control locations (Cape Kennedy offsite sampling sites). Despite differences in potassium-40 levels in soil, levels of potassium-40 in vegetation are not significantly different between Crystal River and comparable control locations.

4. Gross beta radioactivity levels in surface water are highly dependent on the salinity of the water. Comparisons of sampling locations will require normalization to the salinity levels to be significant.

5. A significant difference exists between levels of gross beta radioactivity in fish samples taken at Crystal River and those taken at Turkey Point,

with Crystal River higher. There is no significant difference between gross beta radioactivity levels in oyster samples taken from Crystal River and from Apalachicola Bay.

6. Differences in cesium-137 levels in vegetation samples taken from certain sampling sites are significant when compared to other sampling sites at Crystal River. The levels are not significantly different from levels in vegetation at control locations in other similar geographic areas.

REFERENCES

(*1*) JOHNSON, W. and C. L. NAYFIELD. Elevated levels of cesium-137 in common mushrooms (*Agaricaceae*) with possible relationship to high levels of cesium-137 in whitetail deer, 1968–1969. Radiol Health Data Rep 11:527–531 (October 1970).

(*2*) FREUND, J. E. Modern elementary statistics. Prentice-Hall, Inc. (1960) p. 294.

(*3*) FREUND, J. E. Modern elementary statistics. Prentice-Hall, Inc. (1960) p. 296.

(*4*) POLIKARPOV, G. G. Radioecology of aquatic organisms. Reinhold Book Division, New York (1966) p. 61.

(*5*) FREUND, J. E. Modern elementary statistics. Prentice-Hall, Inc. (1960) p. 260.

Environmental Levels of Radioactivity at Atomic Energy Commission Installations

1. Oak Ridge Area[1] July-December 1969

Union Carbide Nuclear Company
Oak Ridge, Tenn.

Oak Ridge area is a complex made up primarily of the Y-12 Plant, the Oak Ridge National Laboratory (ORNL), and the Oak Ridge Gaseous Diffusion Plant (ORGDP).

Radioactive waste materials arising from the operation of atomic energy installations in the Oak Ridge area are collected, treated, and disposed of according to their physical states. Solid wastes are buried in a Conasauga shale formation which has a marked ability to fix radioactive ma-

[1] Summarized from Environmental Levels of Radioactivity for the Oak Ridge area, (Report Period July–December 1969). Health Physics and Safety Section, Health Physics Division, Oak Ridge National Laboratory.

terials by an ion exchange mechanism. Liquid wastes which contain long-lived fission products are confined in storage tanks or are concentrated by evaporation and disposed of in deep wells. Low-level liquid wastes are discharged, after preliminary treatment, to the surface streams. Air that may become contaminated by radioactive materials is exhausted to the atmosphere from several tall stacks after treatment by means of filters, scrubbers, and/or precipitators.

Air monitoring

Atmospheric contamination by radioactive materials occurring in the general environment of East Tennessee is monitored by two systems of monitoring stations. One system consists of nine stations which encircle the plant areas (figure 1) and provide data for evaluating the impact of all Oak Ridge operations on the immediate environment. A second system consists of eight stations encircling the Oak Ridge area at distances of from 12 to 75 miles (figure 2). This system provides data to aid in evaluating local conditions and to assist in determining the spread or dispersal of contamination should a major incident occur.

Sampling for radioactive particulates is carried out by passing air continuously through a filter paper. Average radioactivity concentrations are presented in tables 1 and 2. Airborne radioactive iodine is monitored in the immediate environment of the plant areas by passing air through a cartridge containing activated charcoal.

During the July–December 1969 surveillance period, 233 samples were collected from perimeter monitoring stations and analyzed for iodine-131. Of these, the maximum concentration detected was 0.15 pCi/m^3, the minimum was less than 0.01 pCi/m^3, and the average was 0.01 pCi/m^3. The radiation protection standard specified in the AEC Manual for iodine-131 in the ambient atmosphere in uncontrolled areas is 100 pCi/m^3.

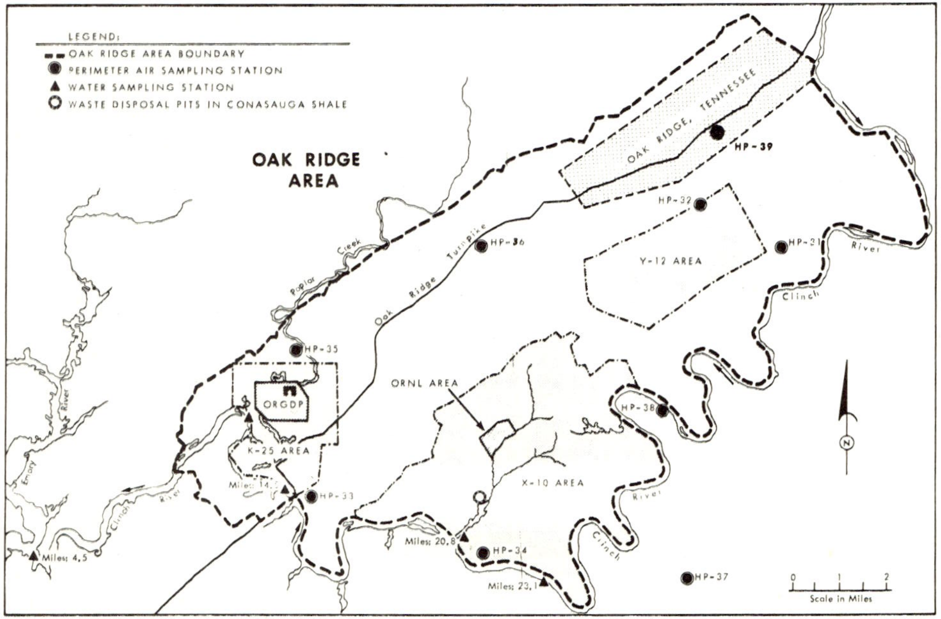

Figure 1. Oak Ridge area, environmental sampling locations

Table 1. Long-lived gross beta radioactivity of particulates in air, Oak Ridge Area, July–December 1969

Perimeter stations						Remote stations					
Station number (figure 1)	Number of samples	Concentration (pCi/m³)			Percent of AEC standard[a]	Station number (figure 2)	Number of samples	Concentration (pCi/m³)			Percent of AEC standard[a]
		Maximum	Minimum	Average				Maximum	Minimum	Average	
HP-31	25	0.43	0.032	0.16	0.16	HP-51	26	0.42	0.047	0.16	0.16
HP-32	26	.53	.025	.19	.19	HP-52	25	.50	.035	.17	.17
HP-33	26	.35	<.001	.12	.12	HP-53	26	.49	:008	.16	.16
HP-34	26	.43	.038	.14	.14	HP-54	25	.42	.035	.15	.15
HP-35	26	.44	.042	.16	.16	HP-55	25	.32	.013	.11	.11
HP-36	[b]119	.43	.019	.18	.18	HP-56	26	.55	.021	.17	.17
HP-37	26	.39	.007	.16	.16	HP-57	26	.46	.038	.16	.16
HP-38	26	.43	.035	.16	.16	HP-58	23	.49	.030	.19	.19
HP-39	26	.57	.040	.19	.19						
Summary	326	0.57	<0.001	0.16	0.16	Summary	202	0.55	0.008	0.16	0.16

[a] The applicable AEC Radiation Protection Standard is 100 pCi/m³.
[b] Sample frequency—5 days per week.

Milk monitoring

Raw milk is monitored for iodine-131 and strontium-90 by the collection and analysis of samples from 12 sampling stations located within a radius of 50 miles of ORNL. Samples are collected weekly at each of eight stations located on the fringe of the Oak Ridge area. Four stations, located more remotely with respect to the Oak Ridge operations, are sampled at a rate of one station each week. The purpose of the milk sampling program is twofold: first, samples collected in the immediate vicinity of the Oak Ridge area provide data by which one may evaluate possible exposure to the neighboring population resulting from waste releases from Oak Ridge operations; second, samples collected at the more remote stations provide background data which are essential in establishing the proper index for the evaluation of data obtained from local samples. The concentrations of iodine-131 and strontium-90 detected in raw milk samples during the period are given in table 3.

Water monitoring

Large volume, low-level liquid wastes originating at ORNL are discharged, after some preliminary treatment, into the Tennessee River system by way of White Oak Creek and the Clinch River. Liquid wastes originating at the ORGDP and Y-12 Plant are discharged to Poplar Creek and thence to the Clinch River. Releases are controlled so that resulting average concentrations in the Clinch River comply with AEC radiation protection standards. The radioactivity concentration from White Oak Creek is measured, and concentration values for the Clinch River are calculated on the basis of the dilution provided by the river.

Table 2. Long-lived gross alpha radioactivity of particulates in air, Oak Ridge Area, July–December 1969

Perimeter stations						Remote stations					
Station number (figure 1)	Number of samples	Concentration (pCi/m^3)			Percent of AEC standard[a]	Station number (figure 2)	Number of samples	Concentration (pCi/m^3)			Percent of AEC standard[a]
		Maximum	Minimum	Average				Maximum	Minimum	Average	
HP-31	25	0.006	<0.001	0.002	0.10	HP-51	26	0.003	<0.001	0.001	0.05
HP-32	26	.008	< .001	.002	.10	HP-52	25	.005	< .001	.002	.10
HP-33	26	.005	< .001	.002	.10	HP-53	26	.004	< .001	.001	.05
HP-34	26	.003	< .001	.001	.05	HP-54	25	.004	< .001	.001	.05
HP-35	26	.008	< .001	.002	.10	HP-55	25	.003	< .001	.001	.05
HP-36	[b]119	.025	< .001	.005	.25	HP-56	26	.004	< .001	.001	.05
HP-37	26	.005	< .001	.002	.10	HP-57	26	.004	< .001	.001	.05
HP-38	26	.003	< .001	.001	.05	HP-58	24	.006	< .001	.002	.10
HP-39	26	.008	< .001	.002	.10						
Summary	326	0.025	<0.001	0.002	0.10	Summary	203	0.006	<0.001	0.001	0.05

[a] The applicable AEC Radiation Protection Standard for natural uranium in air is 2 pCi/m^3.
[b] Sample frequency—5 days per week.

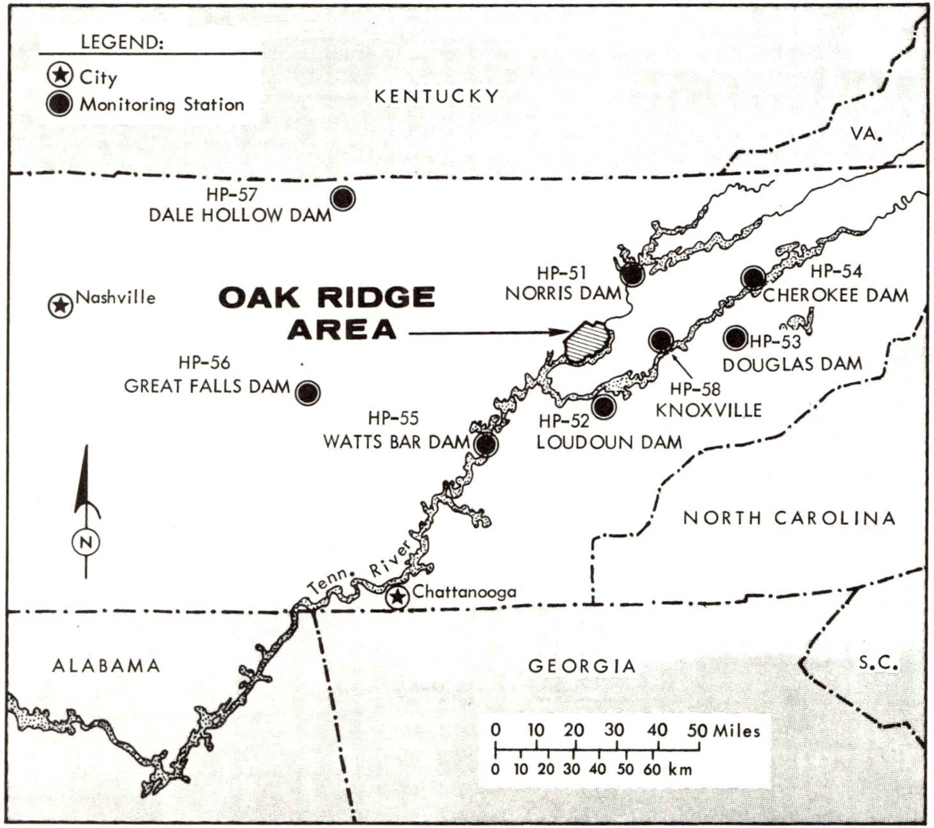

Figure 2. Remote air monitoring stations, Oak Ridge area

Table 3. Radionuclides in raw milk, Oak Ridge area July–December 1969

Radionuclide and location	Concentration (pCi/liter)		
	Maximum	Minimum[a]	Average
Iodine-131			
Immediate environs	48	<10	<10
Remote environs	<10	<10	<10
Strontium-90			
Immediate environs	32	2.0	15
Remote environs	22	5.4	15

[a] The minimum detectable concentrations of iodine-131 and strontium-90 in milk are 10 pCi/liter and 2 pCi/liter, respectively. In averaging, one-half of the minimum detectable concentration was used for all samples showing an activity less than this concentration.

Water samples are taken at a number of locations in the Clinch River, beginning at a point above the entry of wastes into the river and ending at Center's Ferry near Kingston, Tenn. Stream gauging operations are carried on continuously to obtain dilution factors for calculating the probable concentrations of wastes in the river. Samples are analyzed for the long-lived beta-particle emitters, uranium, and the transuranic alpha-particle emitters.

Analyses are made of the effluent for the long-lived radionuclides only, since cooling time and hold-up time in the waste effluent system are such that short-lived radionuclides are normally not present. The average concentrations of major radionuclides in the Clinch River are given in table 4 for the period from July–December 1969.

Two quarterly composite water samples from the Clinch River were analyzed for uranium during this period. For these samples, the uranium concentrations were less than 1 pCi/liter, the level of detectability. The AEC radiation protection standard for natural uranium in water released to unrestricted areas is 20 nCi/liter.

Table 4. Concentrations of major radionuclides in the Clinch River, July–December 1969

Radionuclide	Average concentration (pCi/liter)		
	Location on Clinch River[a]		
	Mile: 23.1 (Upstream)	Mile: 20.8[b] (Outfall)	Mile: 4.5 (Downstream)
Strontium-90	1.3	0.8	0.7
Cerium-144	.2	<.1	<.1
Cesium-137	1.4	.2	1.2
Ruthenium-103-106	.4	.3	1.5
Cobalt-60	1.5	.1	.7
Zirconium-niobium-95	.3	<.1	.5
Tritium	NA	1,740	NA

[a] The location on Clinch River is given in terms of the distance upstream from the Tennessee River (figure 1).

[b] The concentrations at Mile: 20.8 are not measured directly but the values are calculated based on the levels of waste released from White Oak Dam and the dilution provided by the Clinch River.

NA, no analysis.

Gamma measurements

External gamma radiation levels are measured monthly at a number of locations in the Oak Ridge area. Measurements are taken with a Geiger-Mueller tube at a distance of 3 feet above the ground. The results are shown in table 5.

Discussion of data

The average air contamination level for gross beta radioactivity, as shown by the continuous air monitoring filter data, for both the immediate and remote environs of the plants (figures 1 and 2) was 0.16 percent of the AEC radiation protection standard for uncontrolled areas.

The average air contamination levels for gross alpha radioactivity, as shown by the continuous air monitoring filter data, for the immediate and remote environs of the plants were 0.10 percent and 0.05 percent, respectively, of the AEC radiation protection standard for natural uranium for uncontrolled areas.

The average concentration of iodine-131 in air in the immediate environs of the plants was 0.01

Table 5. External gamma radiation levels,[a] Oak Ridge area, July–December 1969

Location	Radiation levels (mR/h)						
	July	August	September	October	November	December	Average
Kerr Hollow Gate	NA	0.012	0.012	0.010	0.012	0.013	0.012
Y-12, East Portal	NA	.011	.013	.011	NA	.018	.013
Newcomb Road	NA	.011	.012	.011	NA	.012	.012
Gallaher Gate	NA	.011	.010	.014	.019	.014	.014
White Wing Gate	NA	.011	.011	.010	.011	.011	.011

[a] The background in the Oak Ridge area, determined in 1943, was approximately 0.012 mR/h.
NA, no analysis.

pCi/m^3. This is approximately 0.01 percent of the AEC radiation standard for uncontrolled areas.

The average concentrations of iodine-131 in raw milk in the immediate and remote environs of the Oak Ridge area were both less than 10 pCi/liter. These levels fall within the limits of FRC Range I if the average intake per individual is assumed to be 1 liter of milk per day.

The average concentration of strontium-90 in raw milk in both the immediate and remote environs of the controlled areas was 15 pCi/liter. The average falls within the limits of FRC Range I for transient rates of daily intake of strontium-90 for application to the average of suitable samples of an exposed population.

The percent maximum concentration of radioactivity in the Clinch River at Mile 20.8 (the point of entry of most of the wastes) and the measured average concentration at Mile 4.5 (near Kingston, Tenn.) were 0.39 percent and 0.25 percent, respectively, of the AEC radiation protection standards. The higher than normal concentration reported for radioactivity at Mile 23.1, Melton Hill Dam, was due primarily to cobalt-60. This contamination was found in the river upstream from the ORNL waste outfall and was the result of a source of contamination other than Oak Ridge operations. The average concentration of transuranic alpha particle emitters in the Clinch River at Mile 20.8 was 0.14 pCi/liter which is 0.14 percent of the AEC radiation protection standard for water containing a mixture of unknown radionuclides.

The average concentration of natural uranium materials in the Clinch River reflecting the effects of all Oak Ridge plants was less than 0.1 percent of the AEC radiation standard for uranium.

The average external gamma radiation measured in the town of Oak Ridge and at the perimeter of the Oak Ridge area was 0.012 mR/h, which is approximately the same as that level measured during the period prior to Oak Ridge operations.

2. Savannah River Plant[2] July-December 1970

E. I. duPont de Nemours
Aiken, S.C.

The Savannah River Plant (SRP), built and operated for the Atomic Energy Commission by E. I. duPont de Nemours and Company, occupies an area of 312 square miles along the Savannah River, 22 miles downstream from Augusta, Ga. Production facilities include a fuel preparation area, three reactors, two fuel separation areas, and a heavy water production plant. A basic goal in plant operation is total containment of radioactive waste. Although some very low level gaseous and liquid wastes are discharged to the environment in controlled releases, dispersal is adequate to ensure environmental concentrations below recommended guides.

A continuous monitoring program has been maintained since 1951 (before plant startup) to determine the concentrations of radioactive materials in a 1,200-square mile area outside the plant. Included in this area are parts of Aiken, Barnwell, and Allendale counties in South Carolina, and Richmond, Burke, and Screven counties in Georgia. This surveillance determines the magnitude and origin of any radioactivity above natural levels. Measured concentrations of radionuclides in air, water, and milk are compared with the AEC radiation protection standards as given in the AEC Manual.

Sensitive instruments, which can detect traces of radioactive materials far below concentrations of hazard significance, are used to determine radioactivity in the environs. Plant-released radioactivity and atmospheric fallout are included in the reported concentrations. Maximum and minimum values given are for individual samples collected during the report period.

[2] Summarized from "Effects of the Savannah River Plant on Environmental Radioactivity," Semiannual Report, July-December 1970 (DPST 10 30 1).

Atmospheric monitoring

Concentrations of radioactive materials in the atmosphere were measured by biweekly analyses of air filters collected at five monitoring stations near the plant perimeter and 10 stations around a circle of about 25 miles radius from the center of the plant (figure 3). Deposition rates of radioactive material at each station were also determined by monthly analyses of rainwater ion exchange columns (fallout collectors). The monitoring stations are spaced so that a significant release of airborne radioactivity by SRP would be detected regardless of the prevailing wind. All stations operate continuously. Four additional air monitoring stations at Savannah and Macon, Ga., and at Columbia and Greenville, S.C., are so distant from SRP that the effect of SRP opera-

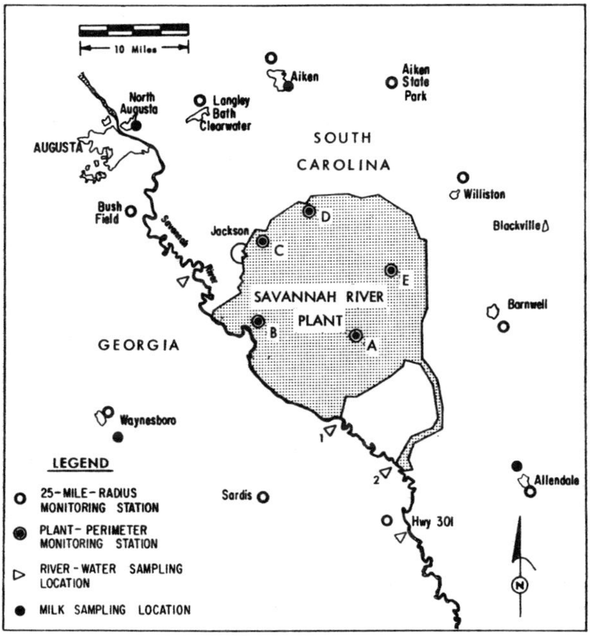

Figure 3. Environmental sampling locations, Savannah River Plant

tions is negligible; they serve as reference points for determining background radioactivity levels (figure 4). This system permits comprehensive surveillance of atmospheric radioactivity and also makes it possible to differentiate between fallout and SRP releases.

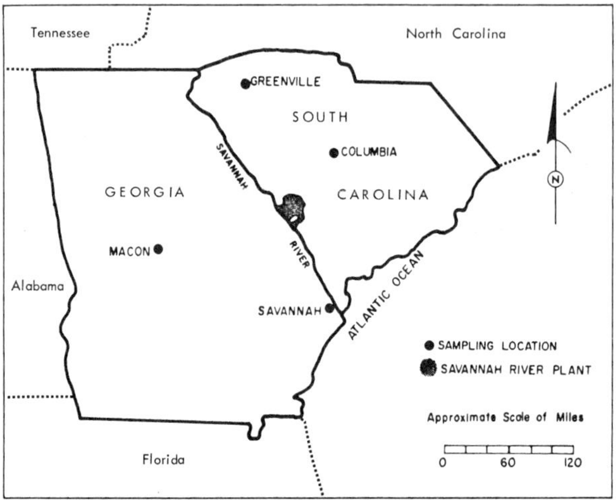

Figure 4. Distant air monitoring stations, Savannah River Plant

The small amount of filterable beta radioactivity released to the atmosphere, primarily from the fuel separation areas, was obscured by fallout. The influence of nuclear tests, which resumed in September 1961, is shown in figure 5. The present low levels of atmospheric activity are attributed to the moratorium on above ground nuclear tests, which began in 1962. The observed spring increases are anticipated and result primarily from the mixing of the stratospheric debris into the troposphere. The average beta concentration in air decreased from 0.26 pCi/m^3 to 0.17 pCi/m^3 during the last half of 1970. Gamma-emitting radionuclides in fallout were cesium-134–137, cerium-141–144, ruthenium-103–106, and zirconium-niobium-95.

Table 6. Radioactivity in air, Savannah River Plant, July–December 1970

Sampling points	Alpha radioactivity[a] (fCi/m³)			Nonvolatile beta radioactivity[b] (pCi/m³)			Specific radionuclides in composite samples (pCi/m³)					
	Maximum	Minimum	Average	Maximum	Minimum	Average	$^{89,90}Sr$	^{137}Cs	$^{141,144}Ce$	$^{103,106}Ru$	^{95}Zr-Nb	^{7}Be
Plant perimeter:												
A	1.3	0.4	0.7	0.51	0.10	0.18						
B	1.5	.4	.8	.66	.08	.14						
C	1.6	.4	.7	.61	.11	.19	0.005	0.01	0.07	0.07	0.14	0.18
D	1.3	.5	.8	.95	.09	.22						
E	1.8	.4	.6	.46	.10	.18						
Average			.7			.18						
25-mile radius:												
Aiken Airport	1.4	.4	.7	.52	.13	.20						
Aiken State Park	1.0	ND	.6	.46	.01	.16						
Allendale	3.0	.3	1.0	.41	.06	.17						
Barnwell	1.2	.3	.7	.46	.07	.18						
Bush Field	1.4	ND	.7	.56	.07	.17	.005	.01	.06	.05	.12	.16
Langley	1.7	.4	.7	.68	.09	.20						
Sardis	1.0	.3	.6	.47	.06	.18						
Waynesboro	1.3	.3	.6	.50	.09	.18						
Williston	1.3	.4	.7	.54	.08	.19						
Highway 301	1.4	ND	.6	.49	.11	.20						
Average			.7			.18						
Distant air-monitoring:												
S.C.												
Columbia	1.4	ND	.9	0.59	.05	.14						
Greenville	1.4	ND	.7	.58	.04	.18						
Ga.												
Macon	1.3	ND	.6	.52	.05	.15	.005	.01	.07	.34	.12	.18
Savannah	1.0	ND	.5	.47	.04	.16						
Average			.7			.16						

[a] AEC radiation protection standard-70 fCi/m³.
[b] AEC radiation protection standard-100 pCi/m³.
ND, nondetectable.

Trace amounts of short radioactive-half-life materials were detected in routine air samples taken with a high volume air sampler. Nonplant origin of the radioactivity was indicated by the uniform levels found at all stations, both on and off the plant. The major component, beryllium-7, is a naturally occurring radionuclide formed by interaction of cosmic rays with oxygen and nitrogen in the upper atmosphere. Radioactivity in air, determined from filter analyses, is shown in table 6. The July–December 1970 concentrations of filterable beta radioactivity (0.17 pCi/m^3) and alpha radioactivity (0.7 fCi/m^3) in air were 0.17 and 1.0 percent of the respective AEC standards. Tritium oxide concentrations in air, at the plant perimeter and at the 25-mile radius stations, did not exceed 0.2 percent of the AEC standard.

Deposition of fallout during July–December 1970 averaged 17 nCi/m^2 at the plant perimeter locations and 15 nCi/m^2 at 25-mile radius locations; comparable values for the last half of 1969 were 13 and 12 nCi/m^2 (excluding naturally occurring beryllium-7). Deposition at each sampling location is presented in table 7.

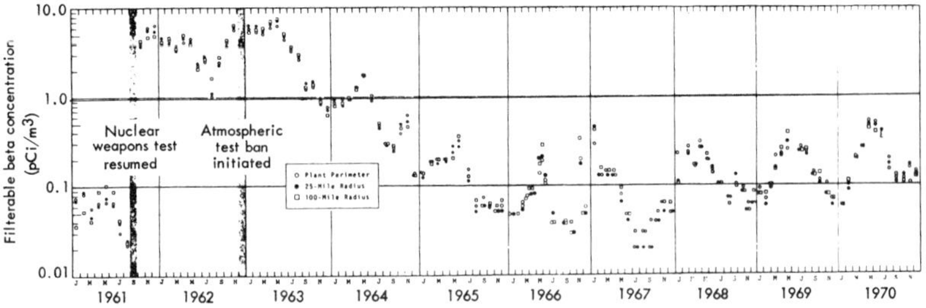

Figure 5. Influence of weapons tests

Table 7. Total fallout deposited, Savannah River Plant, July–December 1970

Sampling points	Radioactivity concentration (nCi/m^2)								
	Alpha[a]	Strontium-89	Strontium-90	Cesium-137	Cerium-141, 144	Zirconium-niobium-95	Beryllium-7[b]	Iodine-131	Ruthenium-103–106
Plant perimeter:									
A	4.4	1.3	0.5	0.8	6.4	4.8	14.7	1.6	1.1
B	2.8	.2	.5	.7	4.9	1.7	14.0	1.0	.8
C	5.0	.6	.7	.8	6.5	4.9	21.5	.7	2.5
D	5.9	.7	.8	.9	8.3	7.6	21.6	1.2	2.6
E	5.0	.9	.8	.9	7.6	5.8	22.5	1.6	1.8
Average	4.6	.8	.7	.8	6.7	5.0	18.9	1.2	1.8
25-mile radius:									
Aiken Airport	1.7	1.1	.5	.8	5.5	3.9	19.8	1.0	1.6
Aiken State Park	3.7	.5	.7	.7	7.5	5.0	21.7	1.6	1.5
Allendale	4.5	.8	.6	.6	7.9	5.2	24.7	1.9	2.1
Barnwell	1.5	.3	.5	.4	1.0	2.7	3.0	ND	1.4
Bush Field	2.3	ND	.7	.8	6.1	5.2	14.5	.8	.8
Langley	4.5	.8	.5	.8	6.2	4.6	19.1	.8	2.0
Sardis	2.6	1.0	.6	1.0	7.1	9.3	19.1	1.0	2.1
Waynesboro	5.5	.7	.7	1.1	7.4	5.6	21.0	.9	1.3
Williston	2.0	.2	.3	.3	5.5	2.2	12.4	1.9	ND
Highway 301	2.4	1.4	.6	.8	4.6	4.6	13.3	1.0	ND
Average	8.0	.7	.6	.7	5.9	4.8	16.9	1.1	1.3

[a] Multiple by 10^{-2}.
[b] A naturally-occurring radionuclide.
ND, less than the sensitivity of analysis.

The plant site is drained by five streams that flow several miles through the reservation before reaching the river (figure 6). The primary sources of the very small amount of radioactivity that reaches the river are the reactor facilities. The reactors are cooled and moderated by heavy water which, in turn, is cooled by river water in heat exchangers. This arrangement prevents irradiation of the river water so that radioactivity is discharged into the river water only on the rare occasions when small quantities of moderator are lost by heat exchanger leaks.

The irradiated fuel (canned to prevent leaching of radionuclides) discharged from a reactor is stored in a large, water-filled basin that is purged to maintain clarity and to control water temperature. Newly discharged fuel is placed into an isolated section of the basin that is equipped with a water circulating system. Portable filters and deionizers are provided to remove a large percentage of the radionuclides, and a heat exchanger is used to control basin water temperature. Tritium from the irradiation of the D_2O moderator accounts for the largest quantity of radioactivity released by the reactors to the environment. However, the contribution to the Savannah River results in concentrations less than 1 percent of the AEC standard.

In January 1965, the Beaufort-Jasper Water Authority began operation of a treatment facility to furnish sanitary water, partially supplied from the Savannah River, to most of Beaufort County, S.C. Water is supplied through a canal from the river at a location about 90 miles below the Savannah River Plant. The city of Savannah also supplements its domestic well water supply with river water during periods of peak demand. The tritium concentrations in raw water collected from the Beaufort-Jasper Water Plant averaged 2.1 nCi/liter (0.07 percent of the AEC standard) during the July–December period. The annual radiation exposure of an individual in this population due to the consumption of 1.2 liters per day of water containing the very low concentration of tritium is 0.19 mrem, (this exposure may be compared with 20 mrem per year from natural potassium found in the body of all persons or with the 170 mrem per year specified by the Federal Radiation Council as the annual dose guide for members of the public).

Communities near the plant get domestic water from deep wells or surface streams. Public water supplies from 14 surrounding towns were collected and analyzed in October. There was no evidence that SRP contributed radioactivity to drinking water supplies; concentrations of alpha radioactivity (1.1 pCi/liter) and beta radioactivity (5 pCi/liter) were essentially the same as those observed before plant startup. Specific analyses showed that the higher than average alpha activity found in some drinking water supplies (for example, 5.5 pCi/liter at Jackson) was due to naturally-occurring radioactivity, primarily thorium-228 and its short-lived daughter products. Radioactivity in public water supplies data is presented in table 8.

Table 8. Radioactivity in public water supplies, July–December 1970

Sampling points	Alpha radioactivity[a] (pCi/liter)	Nonvolatile beta radioactivity[b] (pCi/liter)
Aiken	0.9	ND
Allendale	ND	ND
Augusta	1.0	8
Barnwell	ND	ND
Bath	1.8	ND
Blackville	.4	ND
Clearwater	ND	ND
Jackson	5.5	10
Langley	2.1	4
New Ellenton	1.0	ND
North Augusta	ND	9
Sardis	ND	ND
Waynesboro	ND	ND
Williston	1.5	ND
Average	1.1	5

[a] AEC radiation protection standard—10 pCi/liter; sensitivity of analysis—0.2 pCi/liter.
[b] AEC radiation protection standard—3 nCi/liter; sensitivity of analysis—4.0 pCi/liter.
ND, less than the sensitivity of analysis.

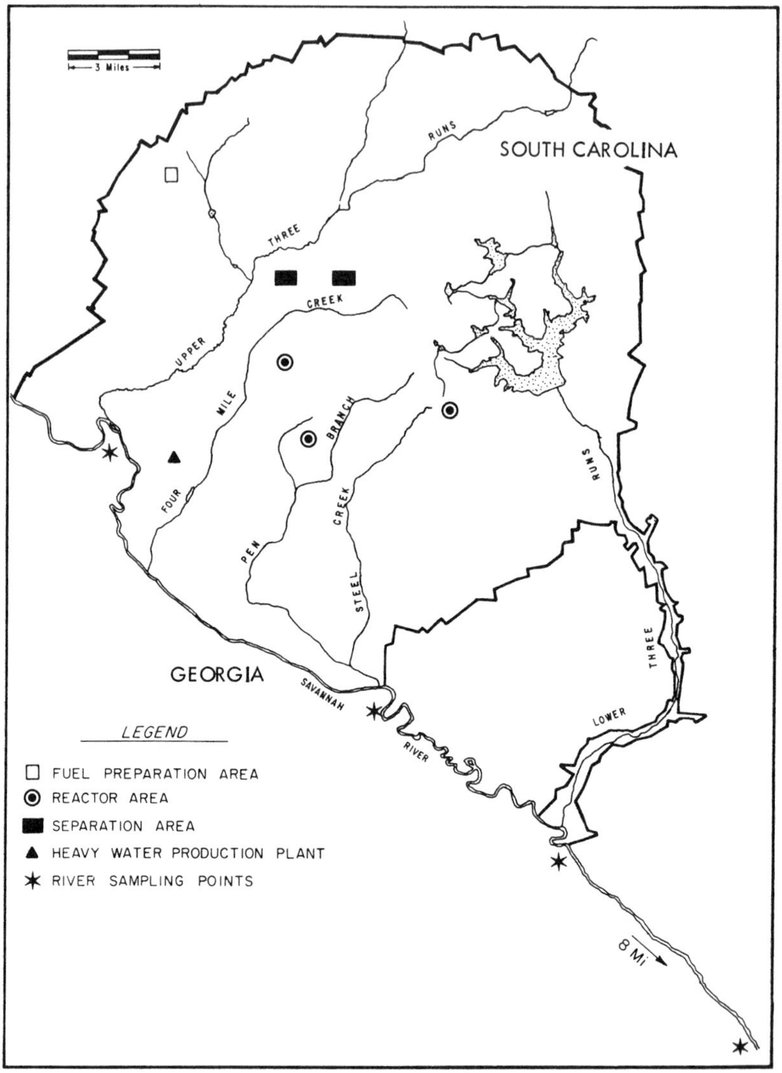

Figure 6. SRP production and effluent streams

River water, analyzed weekly, was sampled continuously at four locations, as shown in figure 6. Concentrations of alpha and nonvolatile beta radioactivity in river water for the past year are presented in table 9. Average concentrations of specific radionuclides found in river water during July–December 1970 appear in table 10.

Tritium, sulfur-35, and trace amounts of cesium-137, strontium-89, and strontium-90 were the radionuclides of SRP origin detectable in river water at the downstream location. Strontium-90 and tritium from worldwide fallout were also detected in river water upstream from SRP effluents.

Tritium, a beta-particle emitter and the most abundant radionuclide released to the river, is produced by neutron irradiation of heavy water moderator in the reactors. The concentration of tritium in river water averaged 0.20 percent of the AEC standard.

Fish in Savannah River

Fish, predominantly bream, were collected weekly upstream, adjacent to, and downstream from Savannah River Plant effluents. Specific radionuclide analyses of fish are summarized in table 11. The radioactivity in bone and flesh from fish collected adjacent to and downstream from SRP indicates some minor contribution by SRP. Although measurably higher than similar collections at the control station upstream from SRP effluents, the increase is of minor significance when referenced against intake guides defined by the AEC radiation protection standard.

Table 9. Radioactivity in Savannah River water, July–December 1970

Sampling point	Alpha radioactivity[a] (pCi/liter)				Nonvolatile beta radioactivity[b] (pCi/liter)			
	July–December 1970			January–June 1970	July–December 1970			January–June 1970
	Maximum	Minimum	Average	Average	Maximum	Minimum	Average	Average
1 mile upstream from Upper Three Runs Creek (control)	0.4	ND	ND	ND	8	ND	4	ND
Downstream from Steel Creek	.5	ND	ND	ND	26	ND	7	4
Downstream from Lower Three Runs Creek	.4	ND	ND	ND	16	ND	8	8
8 miles downstream from Lower Three Runs Creek (Highway 301)	.3	ND	ND	ND	15	ND	7	6

[a] AEC radiation protection standard—10 pCi/liter; sensitivity of analysis—0.2 pCi/liter.
[b] AEC radiation protection standard—3 nCi/liter; sensitivity of analysis—4.0 pCi/liter.
ND, less than the sensitivity of analysis.

Table 10. Average concentration of radionuclides in Savannah River water July–December 1970

Radionuclides	Concentration (pCi/liter)			
	Sensitivity	Control (1 mile upstream from Upper Three Runs Creek)	Highway 301 (8 miles downstream from Lower Three Runs Creek)	Percent AEC standard at Highway 301
Tritium	600	730	5,850	0.20
Sulfur-35	5.0	ND	4.1	.01
Chromium-51	4.3	ND	ND	<.001
Manganese-54	.4	ND	ND	<.0004
Cobalt-60	1.4	ND	ND	<.005
Zinc-65	1.1	ND	ND	<.001
Strontium-89	.3	ND	1.5	.05
Strontium-90	.01	.6	1.0	.33
Zirconium-niobium-95	.5	ND	ND	<.001
Ruthenium-103-106	3.2	ND	ND	<.03
Iodine-131	.2	ND	ND	<.07
Cesium-134-137	.6	ND	.9	.008
Barium-lanthanum-140	1.6	ND	ND	<.01
Cerium-141-144	2.5	ND	ND	<.02
Neptunium-239	2.2	ND	ND	<.002

ND, nondetectable, less than sensitivity of analysis.

Table 11. Radioactivity in Savannah River fish July–December 1970

Location	Type of sample[a]	Concentration (pCi/g wet weight)			
		Strontium-89,-90		Cesium-137	
		Maximum	Average	Maximum	Average
Above plant boundary (control)	Bone	38	22	<2	<2
	Flesh	2	<1	<1	<1
Adjacent to plant	Bone	39	25	<2	<2
	Flesh	2	<1	3	2
Below plant at Highway 301	Bone	25	18	<2	<2
	Flesh	<1	<1	2	2

[a] Monthly composites of weekly collections for strontium-89, -90 and cesium-137.

Vegetation

Radioactive contamination of growing plants may result from sorption of radioactive materials from the soil or from foliar deposition.

Grass samples were collected at seven locations along the plant perimeter and at seven other locations along a 25-mile radius route (these are not designated on figure 7). Bermuda grass was selected for analysis because of its importance as a pasture grass for dairy herds and its availability during all seasons of the year. Samples from each quadrant of the plant site and of the surrounding area were composited for monthly analysis. Gamma-emitting radionuclides in grass samples (excluding beryllium-7) were from fallout. Alpha-particle emitters averaged 0.1 pCi/g at the plant perimeter and at the 25-mile radius locations, the same as observed during the last half of 1969; gamma-ray emitters averaged 11.1 and 9.9 pCi/g, respectively, as compared to 12.0 and 15.3 pCi/g for the first half of 1969. Radioactivity in grass samples is presented in table 12.

Table 12. Radioactivity in vegetation, SRP July–December 1970

Radionuclides	Sensitivity of analysis (pCi/g dry weight)	Concentration (pCi/g dry weight)			
		Plant perimeter (7 locations)		25-mile radius (7 locations)	
		Maximum	Average	Maximum	Average
Alpha emitters	0.10	0.5	0.1	0.6	0.1
Cesium-137	.3	2.0	.7	.9	.5
Cerium-141, 144	1.0	9.9	4.3	5.3	3.6
Ruthenium-103, 106	1.4	ND	ND	ND	ND
Beryllium-7	3.0	10.2	4.8	8.2	4.4
Iodine-131	.2	ND	ND	.9	.2
Barium-lanthanum-140	6.0	ND	ND	ND	ND
Zirconium-niobium-95	.5	1.5	1.3	2.2	1.2

ND, less than sensitivity of analysis.

Milk

Milk was sampled at three dairies within a 25-mile radius of the Savannah River Plant (figure 7). Samples, collected biweekly, were analyzed for tritium and radioiodine. Strontium-90 and cesium-137 determinations were made quarterly. Milk produced in the area and sold by major distributors was also analyzed for these radionuclides. Results from analyzing milk for radioactivity during July–December 1970 are shown in table 13.

Average concentrations of the radionuclides in milk were 12 pCi/liter of strontium-90 and 40 pCi/liter of cesium-137 compared to 11 pCi/liter of strontium-90 and 20 pCi/liter of cesium-137 during the first half of 1970. Iodine-131 was less than the sensitivity of the analysis (5 pCi/liter) throughout this period. These values are consistent with those reported by the U.S. Public Health Service for most sections of the United States. Tritium in local milk, when present, is assumed to be associated with plant operations. The average tritium level (1,012 pCi/liter) was 0.03 percent of the AEC standard for water.

Table 13. Radioactivity in milk from local dairies, Savannah River Plant July–December 1970

Distributing points	Tritium[a] (nCi/liter)			Strontium-90[b] (pCi/liter)		Cesium-137[c] (pCi/liter)	
	Maximum	Minimum	Average	September	December	September	December
Aiken	1.0	ND	0.4	9	11	31	33
North Augusta	3.0	0.9	1.7	19	21	59	66
Waynesboro	2.5	.4	1.1	7	8	34	34
Major distributors[d]	1.7	.4	.8	10	10	36	30

[a] Sensitivity of analysis—0.2 nCi/liter, AEC standard—3,000 nCi/liter.
[b] Sensitivity of analysis—1.0 pCi/liter, AEC standard—300 pCi/liter.
[c] Sensitivity of analysis—20 pCi/liter, AEC standard—20 nCi/liter.
[d] Milk produced in local dairies but sold by major distributors.
ND, nondetectable, less than sensitivity of analysis.

Environmental gamma radiation levels

Monthly measurements of environmental gamma radiation were made with thermoluminescent dosimeters. The July–December 1970 data (table 14) are characteristic of measurements observed at individual stations for the past several years.

Table 14. Environmental gamma radiation, Savannah River Plant, July–December 1970

Sampling points	Gamma radiation (mR/24 h)		
	Maximum	Minimum	Average
Plant perimeter:			
A	0.24	0.16	0.19
B	.17	.13	.15
C	.21	.11	.15
D	.28	.12	.17
E	.16	.11	.14
Average			0.16
25-mile-radius:			
Aiken Airport	0.15	0.13	0.13
Aiken State Park	.16	.11	.13
Allendale	.21	.12	.16
Barnwell	.22	.14	.17
Bush Field	.23	.15	.18
Langley	.19	.13	.16
Sardis	.20	.12	.16
Waynesboro	.27	.10	.17
Williston	.21	.11	.16
Highway 301	.25	.14	.19
Average			0.16

Summary

The quantity of radioactive waste released by the Savannah River Plant to its environs was, for the most part, too small to be distinguished from natural background radiation or was obscured by fallout from offsite sources. Beta radioactivity in air, which had no relationship to plant operations, was about the same as that observed for the same period of 1969 and was 66 percent of the previous 6-month level. Radioactive materials in fish flesh continued to be far below levels considered significant from a health standpoint. The average concentration of any radionuclide in river water at Highway 301 did not exceed 0.3 percent of the AEC radiation protection standard.

The biological effects of radioactive waste

By D. S. Woodhead

1. Introduction

In general, discharges of radioactive waste to the environment are restricted by the public health requirement that the resulting levels of radioactivity in animal or plant species utilized by man, or in that part of the environment accessible to man, should not produce radiation dose rates to members of the general public in excess of the levels recommended by the International Commission on Radiological Protection (Anon. 1966). At the present time the recommended annual dose limit for whole body irradiation is set at 500 mrem a^{-1}, approximately five times the annual dose rate from natural background radiation. For example, in the case of the effluent discharged into the northeast Irish Sea from the U.K.A.E.A. fuel processing factory at Windscale, the discharge rate is limited by the increase in dose rate to the gastro-intestinal tract of that part of the population of South Wales which consumes 'laverbread' (Preston & Jefferies 1967, 1969*a*). This foodstuff is manufactured from seaweeds of the genus *Porphyra* which grow on the beaches in the vicinity of Windscale and which accumulate radioactive ruthenium (principally ^{106}Ru). A secondary consideration in this particular situation concerns the accumulation of radioactivity by silt which is deposited on the mudflats of the Ravenglass estuary to the south of Windscale, and the consequent increased external dose rate of fishermen.

The determination of the limits on any particular discharge to the marine environment only takes account of the possible hazard to man, and the implicit assumption is made that when this hazard is acceptably small the fauna and flora of the marine ecosystem, which are less radiosensitive than man, are not at risk. It is possible, however, that certain components of the marine ecosystem may receive significant radiation doses by virtue of high concentration factors for certain radionuclides, either in the animals themselves or by some part of their habitat, even when the discharges are controlled on the basis of the public health criteria outlined above (Polikarpov 1966).

2. The effects of radiation on marine species

(a) *Acute exposure*

Although the damaging and lethal effects of radiation have been recognized and studied almost since the discovery of X-rays and radioactivity, the total available amount of information relevant to marine organisms is quite small. Nevertheless, it is possible to use the data which are available to reach certain general conclusions.

Table 1. Sensitivities of marine animals to acute irradiation

phylum	species	$LD_{50/30}$, krad	reference
Mollusca	*Nassarius obsoletus*	37	White & Angelovic (1966)
	Mercenaria mercenaria	110	Price (1965)
	Crassostrea virginica	100	Price (1965)
	Urosalpinx cinerea	42	Hargis, Arrighi, Ramsey & Williams (1957)
	Thais lamellosa	*ca.* 20	Bonham & Palumbo (1951)
Arthropoda	*Calliopus laeviusculus*, *Allorchestes angustus*, Dana	12	Bonham & Palumbo (1951)
	Production of young amphipods reduced to 50 % of the control	0.6	Bonham & Palumbo (1951)
	Callinectes sapidus	51	Engel (1967)
	Palaemonetes pugio	1.5	Rees (1962)
	Uca pugnax	8.2	Rees (1962)
Echinodermata	*Arbacia punctulata*	39	White & Angelovic (1966)
Chordata	*Fundulus heteroclitus*	1.1	White & Angelovic (1966)
	Mugil cephalus	1.5	White & Angelovic (1966)
	Eucinostomus sp.	3.5	White & Angelovic (1966)
	Lagodon rhomboides	3.0	White & Angelovic (1966)
	Micropogon undulatus	1.1	White & Angelovic (1966)
	Paralicthys lethostigma	5.6	White & Angelovic (1966)
	Developing plaice (*Pleuronectes platessa*) eggs at blastoderm stage	0.1	Ward, Beach & Dyson (1970)

Table 1 gives data on the tolerance of several marine species to acute whole body irradiation in terms of the median lethal dose at 30 days ($LD_{50/30}$). In broad agreement with the greater amount of information available relating to terrestrial animals it is apparent that the radiosensitivity tends to increase with the increasing biological complexity of the organism.

The high values of $LD_{50/30}$ render it extremely unlikely that acute lethality will ever become a problem in environments contaminated by the controlled disposal of radioactive waste. Nevertheless, two of the results given in table 1 indicate that there are probably much more sensitive criteria which can be used to seek evidence of damage. The low dose (as compared with the $LD_{50/30}$ for adults) required to reduce the number of young amphipods produced to 50 % of control values is indicative of the sensitivity of the gonad to ionizing radiation, and the low LD_{50} at metamorphosis for the developing plaice embryo demonstrates the susceptibility of the embryo to radiation damage.

These results are supported by data obtained from experiments carried out with rainbow trout, in which it was found that exposure of partially mature gametes within the parent fish to approximately 250 R of X-rays resulted in a cumulative mortality of 50 % at hatching, whereas the $LD_{50/30}$ for adult fish was found to be > 2500 R (Foster *et al.* 1949; Welander *et al.* 1949), and experiments carried out with silver salmon eggs, where a particular stage in the first cleavage division of the zygote was found to have an $LD_{50/150}$ of 16 R, and the embryo became more radioresistant with increasing development (Bonham & Welander 1963).

In a contaminated environment ionizing radiation is not the only factor affecting the survival of the animals, and the interactions of acute irradiation, temperature and salinity have been investigated for the estuarine fish *Fundulus heteroclitus* (Angelovic, White & Davis 1968). Table 2 gives some of the results obtained.

TABLE 2. MEDIAN LETHAL DOSES (rad) FOR *FUNDULUS HETEROCLITUS* AS A FUNCTION OF TIME, TEMPERATURE AND SALINITY

number of days after irradiation	salinity ‰	temperature/°C 12	17	22	27
30	5	> 2500	> 2500	2050	1590
	15	> 2500	> 2500	1540	770
	25	> 2500	> 2500	1020	720
60	5	1260	1960	1580	350
	15	1390	1715	1050	325
	25	1890	1500	480	300

At 30 days after irradiation the fish were more tolerant of the radiation insult at low salinities and at the lower temperatures. At 60 days after irradiation the apparent protective effect of low salinity had disappeared at 27 °C, although it was still present at 22 and 17 °C; at 12 °C it was the high salinity environment which appeared to provide protection against the effects of radiation. Thus the water temperature and salinity interact to modify the effects of radiation. The LD_{50} decreases with increasing time after irradiation up to 60 days independently of the temperature and salinity, and also decreases with increasing temperature above 17 °C. It was concluded that the addition of a radiation stress to those naturally present in the environment would decrease the ability of the fish to survive by reducing their tolerance to these stresses.

(b) *Chronic exposure*

From the point of view of possible biological damage in contaminated environments it is more pertinent to consider the effects of chronic irradiation, and again the total amount of information relating to marine species is fairly limited.

The effect of continuous irradiation on the survival and growth of young blue crabs (*Callinectes sapidus*) has been investigated by Engel (1967). The crabs were irradiated by a cobalt-60 γ-ray source at dose rates of 3.2, 7.3 and 28.9 rad h^{-1} for 70 days. At the two lower dose rates the mortality over the term of the experiment was not significantly different from that of the controls, but at the highest dose rate a significant proportion of the total mortality was attributable to the effects of irradiation. All the crabs moulted at least once but the incidence of second and third moults was dependent on dose rate, with the lowest frequency at the highest dose rate. The growth rate was diminished at the highest dose rate, but at the lowest dose rate the growth rate was significantly greater than in the other two irradiated groups and the controls.

The interactions of chronic irradiation, temperature and salinity on the morphology of young pinfish (*Lagodon rhomboides*) have been investigated by White (1969). An analysis of variance was carried out for each of nine morphological parameters after the fish had been maintained for 45 days at nine combinations of the three environmental variables (i.e. temperature: 15, 20, 25 °C; salinity: 10, 20, 30‰; irradiation: 0, 0.83, 1.28 rad h^{-1}). Of the nine parameters, all were significantly altered by temperature, five by salinity, and only two by irradiation. The interactions between irradiation and salinity and irradiation and temperature affected four and eight of the parameters respectively; the interaction between temperature and salinity did not contribute significantly to the variation of any parameter during the period of the experiment. The three-way interaction between irradiation, temperature and salinity affected seven of the nine parameters. It was concluded that the effects of increased radiation dose rate due to waste disposal cannot be considered in isolation, and that as many environmental parameters as possible must be studied simultaneously if the ecological effects of natural and man-made variations are to be properly understood.

(i) *Effects of chronic irradiation on developing fish eggs*

The effects of chronic irradiation on developing fish eggs have been quite extensively studied, although the accumulated results are rather conflicting.

Polikarpov and his co-workers have investigated the effects of low concentrations of ^{90}Sr–^{90}Y on the developing eggs of several Black Sea fishes (Polikarpov 1966). An excess of abnormal larvae at hatching was found at water concentrations of 10^{-10} Ci l^{-1} and greater. The concentration–responsive curve is given in figure 1 and is remarkable in that an increase in the concentration of radioactivity by six orders of magnitude (10^{-10} to 10^{-4} Ci l^{-1}) only increases the incidence of abnormalities at hatching by a factor of four. Similar experiments with anchovy eggs (Polikarpov & Ivanov 1961) showed a two- to threefold increase in egg mortality when the concentration of radioactivity was increased over the range 2×10^{-8} to 2×10^{-4} Ci l^{-1}. On the assumption that the radiation dose to the developing embryo is proportional to the concentration of radioactivity in the water, a more marked dose–effect response might be expected. In general it is found that a 100 % response is produced by a radiation dose which is, at most, 100 times that required to give the minimum detectable response (Templeton, Nakatani & Held 1970). Fedorov, Podymachin, Schitenko & Chumachenko (1964) demonstrated increased mortality and abnormality rates for plaice eggs reared in seawater contaminated with ^{137}Cs and ^{90}Sr–^{90}Y at levels of 10^{-11} and 10^{-10} Ci l^{-1}, whereas Templeton (1966) found no consistently significant effects at ^{90}Sr–^{90}Y levels up to 10^{-4} Ci l^{-1}. Templeton also failed to demonstrate any effects attributable to radiation doses of up to 500 rad delivered chronically during embryonic development.

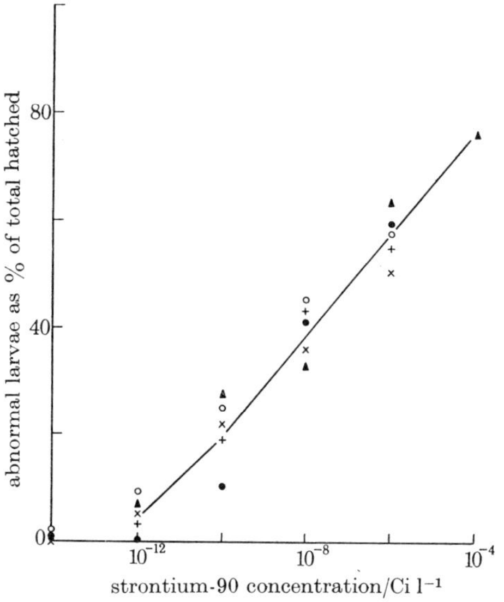

FIGURE 1. The effect of strontium-90 contamination on developing fish eggs. +, mullet; ▲, green wrasse; ○, horse mackerel; ×, anchovy; ●, mixture of pelagic eggs. (From Polikarpov 1966.)

Ivanov (1967) sought more basic indicators of damage in developing embryos and investigated the mitotic activity and the production of chromosome aberrations in the embryos of the Black Sea scorpionfish when the eggs were maintained in water contaminated with ^{90}Sr–^{90}Y. The mitotic activity was found to decrease as the concentration of activity increased, while the incidence of chromosome damage increased, both effects becoming significant at an activity level of 10^{-9} Ci l^{-1}.

Long-term experiments have been carried out at the University of Washington to investigate the effects of chronic exposure of chinook salmon eggs and alevins to γ-irradiation (Donaldson & Bonham 1970). The embryos and larvae were irradiated at 0.5 R d^{-1} from fertilization until absorption of the yolk sac was complete. The total dose thus received was estimated to be forty times that received by the wild population in the contaminated environment of the Columbia River during embryonic and larval development and the migration to the sea. The irradiated fingerlings were fin-clipped for identification, together with a control group, and released to migrate to the sea at 190 days after hatching. At the time of release the irradiated fish were significantly heavier than the controls, but during the following 4 years the numbers of fish from the two groups returning to spawn were not significantly different.

The mature fish from this experiment which returned over the next 4 years were used in a further series of experiments in an attempt to detect the effects of the original radiation insult.

Precocious males of both groups returning after one year were mated with older unirradiated females, the offspring reared to migrating size, marked and released. At release there were no significant differences in the characteristics of the progeny of irradiated and control males, and although more fish of irradiated parentage returned over the next 4 years the difference was not significant.

Fish from both groups returning after 2 years were mated with fish of like irradiation history, the young reared to migrating size, marked, and released. There were no significant differences in the characteristics of the young from the two groups at release, and again, although more fish of irradiated parentage returned during the next 4 years, the difference did not attain significance.

Fish from both groups returning after 3 years were again mated with fish of like irradiation history. The fertile eggs from irradiated parents were split into two groups: one group was irradiated at 0.5 R d^{-1} until the yolk sac had been absorbed and the second was maintained as a control; eggs from unirradiated parents were also reared for control purposes. The fingerlings were marked and released, and there were no significant differences in the returns from the three groups during the following 3 years.

These experiments are continuing with higher dose rates of 1.3, 2.8, 5.0 and 10.0 R d^{-1} given during embryonic and larval development. At the lower two dose rates no significant differences have been found between irradiated and control groups at release.

3. Dose rates in contaminated environments

In the experiments described in § 2 high doses and dose rates were necessary to produce effects, and to assess the likely effects of environmental contamination with radioactivity, estimates of the dose rate due to the contamination are required. As has been pointed out, the highest dose rates will be received by animals which either accumulate large quantities of radioactivity directly from the water or via a food chain, or which live in a habitat that has a high concentration factor for radioactivity, for example benthic animals living in a muddy seabed.

A considerable amount of work has been done to determine the concentration factors for various marine species for the radionuclides likely to be introduced into the environment as a result of the development of the nuclear power industry. Table 3 gives order of magnitude values for the concentration factors for silt, algae and representative groups of animals, for the radionuclides which are of importance in relation to the wastes discharged into the coastal waters of the British Isles (Freke 1967; Lowman, Folsom, Rice & Richards 1970; Preston & Jefferies 1969*b*).

TABLE 3. CONCENTRATION FACTORS FOR MARINE MATERIALS

	radionuclide	silt	Algae	molluscs	Crustacea	fish
fission products	^{137}Cs	5×10^2	10	10^2	10^2	50
	^{144}Ce	10^4	10^3	10^3	10^3	10
	^{90}Sr	1	1	10	10	1
	^{106}Ru	10^4	2×10^3	2×10^3	2×10^2	1
	^{95}Zr–^{95}Nb	10^4	5×10^2	10^3	2×10^2	1
activation products	^{32}P	10^4	10^4	10^4	10^4	2×10^4
	^{35}S	1	1	1	1	1
	^{60}Co	10^4	10^3	10^3	10^3	10^2
	^{65}Zn	10^4	10^3	10^5	4×10^3	2×10^3

These concentration factors can be used to calculate estimates of the mean whole-body dose rates to the animals on the basis of the following assumptions:

(1) that the radioactivity is uniformly distributed throughout the animal; this will obviously lead to underestimates of the dose rates to particular organs where there is preferential accumulation, for example strontium in bone;

(2) that an equilibrium level of radioactivity has been reached in the animal which is appropriate to the maximum levels of radioactivity that have been measured in the water in the environment, i.e. at Windscale in the case of the fission product nuclides and at Bradwell in the case of the activation product nuclides;

(3) that in the case of β-emitting nuclides, the β-radiation is completely absorbed within the animal.

The second and third assumptions are conservative in the sense that they lead to overestimates of the dose rate. The values thus obtained are given in table 4.

To the dose rate from internal contamination must be added that from the external environment. In the case of free-swimming animals this means the negligible contribution from the activity in the water, but in the case of benthic organisms living on or in the bottom it results in a substantial additional dose rate from the radioactivity adsorbed on to the silt. This can best be illustrated by reference to the situation at Windscale, where the nuclides ^{144}Ce, ^{106}Ru and ^{95}Zr–^{95}Nb are heavily concentrated on the seabed close to the outfall, resulting in relatively high dose rates to benthic animals.

Table 4. Maximum dose rates to marine animals in contaminated areas due to the accumulation of radioactivity

radionuclide	dose rate/μrad h^{-1} molluscs	Crustacea	fish
^{137}Cs	5.3	5.9	3.3
^{144}Ce–^{144}Pr	170	170	1.7
^{90}Sr–^{90}Y	1.2	1.2	0.1
^{106}Ru–^{106}Rh	1500	150	0.73
^{95}Zr–^{95}Nb	260	80	0.49
^{32}P	1.8	1.8	3.6
^{35}S	—	—	—
^{60}Co	0.25	0.4	0.05
^{65}Zn	12	0.5	0.3

Table 5. Radiation dose to plaice in the northeast Irish Sea

				dose rate/μrad h^{-1}
background sources of radiation dose	external	natural activity in water		0.2
		natural activity in seabed		6.0
		cosmic radiation at 18 m		0.6
	internal	potassium-40		1.6
			total	8.4
dose rate to plaice from fission nuclides 4 km north of effluent outfall	external	seawater	β	0.1
			γ	0.2
		seabed	β	4000.0
			γ	1890.0
	internal		β	0.3
			γ	0.1
			total	5890.7

Table 5 gives the components of the total dose to plaice, broken down by source and type of radiation. At the particular site chosen, a silty area 4 km north of the effluent outfall, the dose rate from the contamination is predominantly due to the activity in the seabed and is approximately 700 times the dose rate from the natural background. Figure 2 shows schematically the dose rate contours at the seabed–seawater interface in the northeast Irish Sea, as calculated from the measured radioactivity of bottom samples collected in September 1968 on a grid of 26 stations. The actual dose received by an animal in this environment is very dependent on its size, since approximately 80% of the total dose is due to β-radiation which is strongly attenuated by even a few millimetres of tissue. Thus a large animal resting on the bottom will receive mainly a high dose to the under surface, with the dose due to the γ-radiation (the remaining 20% of the total) being fairly uniform throughout the body. A small animal buried in the silt, however, will receive a β-ray dose from all directions and the total dose received will be approximately 1.8 times that given in figure 2. (Again the size of the animal will determine the variation in the magnitude of the β-ray dose within the animal.)

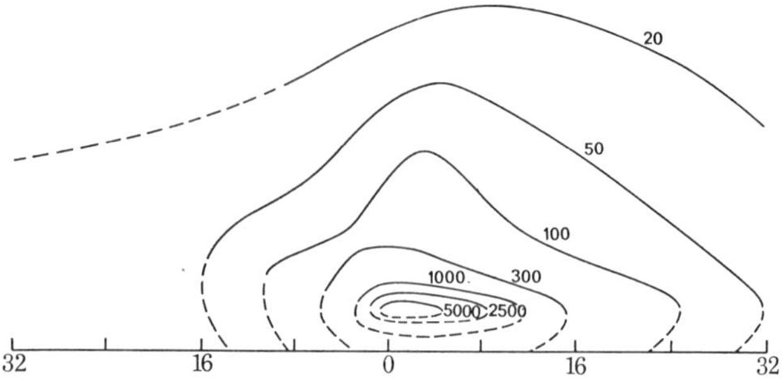

FIGURE 2. Seabed dose rate contours in the vicinity of Windscale. Dose rate in μrad h^{-1}.

A small mollusc, for example *Nucula turgida* or *Cultellus pellucidus*, in the silt close to the pipeline outfall thus receives a dose rate of approximately 18 mrad h^{-1} from the silt and approximately 2 mrad h^{-1} from internal contamination, giving a total of 20 mrad h^{-1} which can be compared with the natural background dose rate of 5–10 μrad h^{-1}. Although the dose rate from the contaminant activity is considerably more than the natural background dose rate, resulting in annual doses of the order of 180 rad, the very high doses necessary to produce damage in irradiation experiments (see table 1) indicate that negligible damage is likely to occur in contaminated environments.

(*a*) In situ *measurements of the dose rate to plaice in the northeast Irish Sea*

The magnitude of the calculated dose rate at the seabed–seawater interface, as given in figure 2, led to the conclusion that a small thermoluminescent dosimeter could be developed which would allow *in situ* measurements of the dose rate received by plaice. The dosimeter consists of 33 mg of thermoluminescent lithium fluoride in a small PVC sachet, and is attached to the shoulder of the fish by means of a Petersen disk tag. The components of the dosimeter and tag are shown in figure 3, together with a marked fish.

One dosimeter was attached on the upper (right) side of the fish and a second on the lower (left) side; in this way it was hoped to obtain an estimate of both the γ-ray and β-ray dose rates and hence a fairly accurate estimate of the dose rate to the gonad. Two releases of marked fish have been made, numbering 2500 in June 1967 and 1100 in April 1968; the overall recapture rate has reached 25 %, from normal commercial fishing operations. The dose rates as indicated by the dosimeters on the underside of the fish are shown in figure 4. The mean dose rate is

360 μrad h^{-1} with an occasional dosimeter indicating a dose rate up to 2.1 mrad h^{-1}. On the basis of the dose rate indicated by the upper dosimeter the mean dose rate to the gonad and whole body is estimated to be 300 μrad h^{-1}.

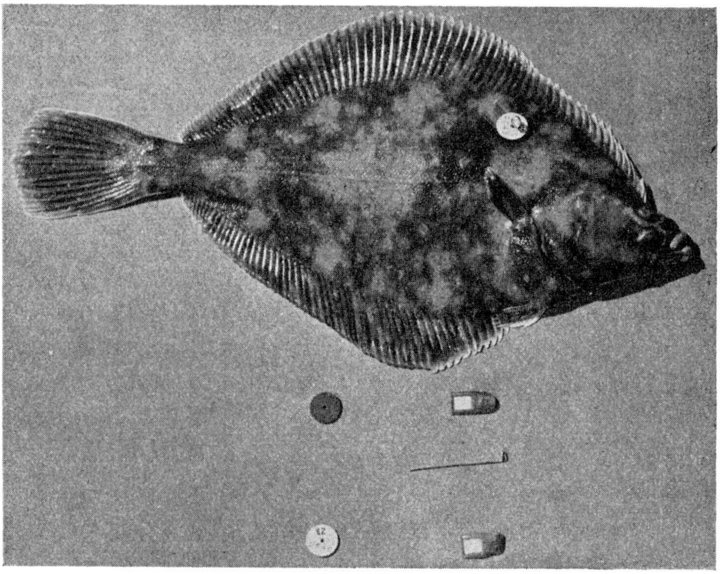

FIGURE 3. Components of dosimeter and Petersen disk tag, together with a marked plaice.

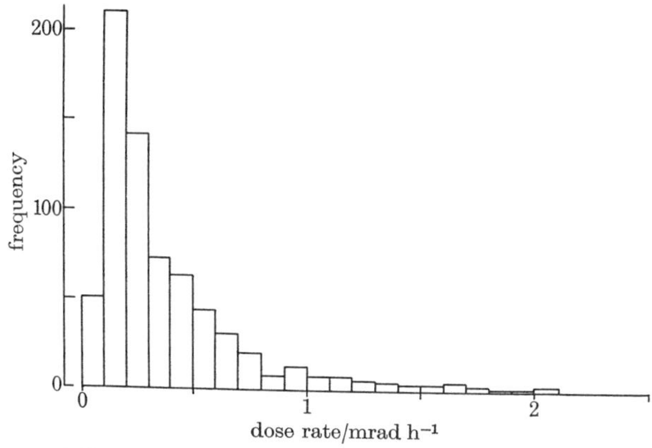

FIGURE 4. Dose rate to plaice in the northeast Irish Sea as indicated by the dosimeter on the underside of the fish.

These results indicate that the natural movement of the plaice reduces considerably the mean dose rate experienced. The mean annual dose is approximately 2.6 rad and is much less than the doses required to produce radiation damage in fish in laboratory experiments. Thus it is extremely unlikely that any somatic effects attributable to the contamination will ever be found.

(b) Estimates of the dose rate to developing plaice embryos in the northeast Irish Sea

The experimental results described in § 2 indicate that the developing embryo is probably more sensitive than the adult animal. Where experiments have been conducted by rearing eggs in contaminated seawater the excess abnormalities or mortalities have generally been attributed to the increased dose rate received by the embryo subsequent to the accumulation of radioactivity by the egg (Polikarpov & Ivanov 1962; Ivanov 1965; Polikarpov 1966). However, estimates of the increased dose rate are rarely given, and the experiment results are conflicting.

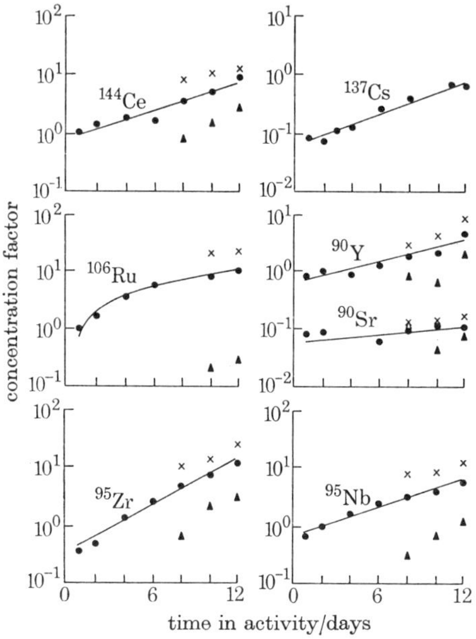

FIGURE 5. The uptake of radioactivity by plaice eggs. ●, concentration factor for whole eggs; ×, concentration factor for egg shells; ▲, concentration factor for egg contents.

Investigations of the uptake of radioactivity by fish eggs have mainly involved the fission product nuclides ^{137}Cs, ^{90}Sr–^{90}Y, ^{106}Ru, ^{144}Ce and ^{95}Zr–^{95}Nb. It is found that both ^{137}Cs and ^{90}Sr, which exist mainly in ionic form in sea water (Revelle & Schaefer 1957), are not accumulated to any great extent by fish eggs, the concentration factor often being less than one. Also, for these two nuclides, the activity is found to be fairly uniformly distributed throughout the egg. For the other nuclides, which exist mainly in particulate form in sea water, much higher concentration factors are found and the majority of the activity is associated with the egg membrane (Polikarpov & Ivanov 1962; Ivanov 1965; Brown & Templeton 1964; Woodhead 1970). Figure 5 shows the uptake curves obtained for plaice eggs under carefully controlled conditions. It can be seen that except for ^{137}Cs and ^{90}Sr the concentration factor reached unity within one or two days and thereafter increased exponentially until the eggs hatched (Woodhead 1970).

Several models have been developed which can be used to calculate the dose rate to the developing embryo if the distribution of activity on and in the egg is known (Fedorov 1965; Adams 1968; Woodhead 1970). The dose rate to developing plaice embryos in a contaminated breeding ground in the northeast Irish Sea has been estimated on the basis of the uptake data given in figure 5 and the results are given in table 6.

TABLE 6. ESTIMATES OF THE DOSE RATE TO DEVELOPING PLAICE EMBRYOS IN THE NORTHEAST IRISH SEA

radionuclide	dose rate from activity on the egg nrad h^{-1}	dose rate from activity in the water nrad h^{-1}	total dose rate from nuclide nrad h^{-1}
^{144}Ce–^{144}Pr	6.5	10.9	17.4
^{137}Cs	1.7	16.0	17.7
^{106}Ru–^{106}Rh	4.7	9.7	14.4
^{90}Sr–^{90}Y	3.7	10.8	14.5
^{95}Zr–^{95}Nb	4.5	16.4	20.9
			84.9
^{40}K	398	298	696

It can be seen that the total dose rate from contaminant radioactivity is an order of magnitude less than that from the natural potassium-40 in the water and the egg. It is difficult to conceive a mechanism whereby this small increment in radiation dose rate could have any adverse effect on the developing plaice embryos.

4. RADIATION EFFECTS ON POPULATIONS

As far as organisms apart from man are concerned it is the total population which is considered to be important rather than the individual. The experimental results given in § 2 indicate that the gonad and the developing embryo are probably more sensitive to radiation than the adult animal; thus it seems that the effects of irradiation are most likely to become manifest as a reduction in the fertility and fecundity of the population.

There are no experimental data available on the effects of irradiation on marine populations, but some investigations have been carried out with freshwater aquatic organisms.

Marshall (1962, 1966) has investigated the effects of chronic irradiation on populations of *Daphnia pulex*. In the first series of experiments the populations were irradiated at exposure rates of between 25 and 75 R h^{-1} and were given a constant *per capita* supply of food. The intrinsic rate of natural increase, r, defined by

$N_t = N_0 e^{rt}$, where N_0 is the population size at time zero and N_t is the population size at time t, was the parameter investigated. It was found that the value of r decreased non-linearly with increasing dose rate, and that the decrease was almost entirely due to a fall in the birth rate; the increase in the death rate did not become a significant factor except at the highest dose rates. The dose rate at which r became zero, i.e. the maximum indefinitely tolerable dose rate, was 70 R h^{-1}.

In the second series of experiments, at dose rates up to 25 R h^{-1}, each population was given a fixed quantity of food regardless of the total number of individuals. Under these more realistic conditions, each population was experiencing a second stress in the form of intra-specific competition for food. It was found that the maximum indefinitely tolerable dose rate was 24 R h^{-1}, a value much lower than that obtained in the first series of experiments and attributable to the extra stress on the populations. In contrast to the result obtained in the first series of experiments it was found that the birth rate increased with increasing dose rate, and it was concluded that the expected reduction in fecundity due to irradiation was over-compensated by the higher *per capita* food supply at the higher dose rates where the mean population size was lower. The most remarkable result obtained from this series of experiments related to the net biomass production, which was found to be independent of radiation dose rate in all the populations that did not become extinct (i.e. at all dose rates up to 24 R h^{-1}).

Blaylock (1969) has investigated the fecundity of a natural population of *Gambusia affinis affinis* (mosquito fish) in an environment contaminated with radioactive waste. The dose rate to the fish was calculated to be 10.9 rad d^{-1} and the fish had been exposed to chronic irradiation for many generations. It was found that the fecundity of the irradiated population was significantly greater than that of a control population from a nearby uncontaminated lake. It was suggested that the increased fecundity might be an adjustment by the population in response to the increased radiation stress. The incidence of dead and abnormal embryos was, however, found to be significantly greater in the irradiated population.

(a) *Genetic effects of radiation*

It is usually assumed that there is no safe radiation dose from the point of view of genetic damage; i.e. any dose of radiation can induce mutations and mutations are almost always harmful. Nevertheless, at the population level with which we are concerned in the marine environment, genetic damage is reparable by the process of natural selection. Thus it seems likely that increased mortality and reduced fertility will be the limiting factors in the recovery of a population from irradiation (Purdom 1966).

The relative unimportance of the genetic effects of irradiation is supported by a calculation of the mutation rate in plaice in the northeast Irish Sea. Schröder (1969) has given values for the radiation-induced mutation rate per gamete in *Lebistes reticulatus* in the range 0.4 to 11×10^{-5} R^{-1}. If it is assumed that a value of 10^{-5} R^{-1} is applicable to plaice in the northeast Irish Sea, where the mean lifetime dose is of the order of 16 rad, then the mutation rate is 1.6×10^{-4} per gamete. If it is assumed that the mutations are dominant lethals resulting in non-viable zygotes then the percentage of dead eggs produced is $3.2 \times 10^{-2}\%$. As Garrod (1966) has emphasized, in a species such as the plaice, which has a relatively high fecundity, the production of young fish far exceeds the resources of the environment to support them, and the superfluous eggs and larvae are reduced by density-dependent mortality, with overall survival rates of 1 % or less. It is evident therefore that the increased mortality at the egg stage due to irradiation will not affect the overall recruitment of fish to the population. It should be made clear that this calculation is valid for only a small proportion of the total Irish Sea plaice stock; in terms of the total population the effect of radiation must be entirely negligible.

Conclusions

It seems fairly clear from the foregoing discussion that the controlled disposal of radioactive waste to the marine environment, as it is conducted at the present time, will not lead to dose rates to the marine biota sufficient to produce detrimental effects. Nevertheless, there are perhaps two areas where further research is required. First, more sensitive indicators of radiation damage are necessary to provide early warning of such damage, and, secondly, investigations are required into the possible synergistic effects of the many pollutants which are discharged to the marine environment.

References (Woodhead)

Adams, N. 1968 Dose rate distributions from spherical and spherical-shell radiation sources, with special reference to fish eggs in radioactive media. *U.K.A.E.A. report AHSB(RP) R* 87, London: H.M.S.O.

Angelovic, J. W., White, J. C. & Davis, E. M. 1968 Annual report of the Bureau of Commercial Fisheries Radiobiological Laboratory, Beaufort N.C. for the fiscal year ending June 30 1967. *U.S. Dept. Int., U.S. Fish and Wildl. Serv., Bur. Comm. Fish. Circ.* no. 289, 35–41.

Anonymous 1966 *Recommendations of the International Commission on Radiological Protection* (adopted 17 September 1965). I.C.R.P. Publication 9. London: Pergamon Press.

Blaylock, B. G. 1969 *Radiat. Res.* **37**, 108–117.

Bonham, K. & Palumbo, R. F. 1951 *Growth* **15**, 155–188.

Bonham, K. & Welander, A. D. 1963 in *Radioecology: proceedings of the first national symposium on radioecology.* (ed. V. Schultz and A. W. Klement), pp. 353–358. Reinhold Publishing Corporation, New York and the American Institute of Biological Sciences, Washington D.C.

Brown, V. M. & Templeton, W. L. 1964 *Nature, Lond.* **203**, 1257–1259.

Donaldson, L. R. & Bonham, K. 1970 *Trans. Am. Fish. Soc.* **99**, 112.

Engel, D. W. 1967 *Radiat. Res.* **32**, 685–691.

Fedorov, A. F., Podymachin, V. N., Schitenko, L. D. & Chumachenko, V. V. 1964 *Voprosy Ikhtiologii* **4**, 579–585.

Fedorov, A. F. 1965 Radiation doses for some types of marine biota under present-day conditions. *Bull. Inst. Océanogr. Monaco* **64**, no. 1334 (*Radioactivity in the sea*, I.A.E.A. publication no. 14).

Foster, R. F., Donaldson, L. R., Welander, A. D., Bonham, K. & Seymour, A. H. 1949 *Growth* **13**, 119–142.

Freke, A. M. 1967 *Hlth Phys.* **13**, 743.
Garrod, D. 1966 *Disposal of radioactive wastes into seas, oceans and surface waters*, p. 859. Vienna: I.A.E.A.
Hargis, W. J., Arrighi, M. F., Ramsey, R. W. & Williams, R. 1957 *Proc. Nat. Shellfish Ass.* **47**, 68–72.
Ivanov, V. N. 1965 *Radiobiology* **5**, 92 (A.E.C.-tr-6598).
Ivanov, V. N. 1967 *Problems of biological oceanography* (trans. 2nd Int. Oceanogr. Congr. 30 May–9 June 1966). Acad. Sci. Ukr. S.S.R., Inst. Biol. Sthn Seas. A. O. Kovalevskii. Kiev: Naukova Dumka, (A.E.C.-tr-6940 1968).
Lowman, F. D., Folsom, T. R., Rice, T. R. & Richards, F. 1970 Accumulation and redistribution of radionuclides by marine organisms. Section VI of *Radioactivity in the marine environment*. National Academy of Sciences—National Research Council, Washington, D.C. (to be published).
Marshall, J. S. 1962 *Ecology* **43**, 598–607.
Marshall, J. S. 1966 *Ecology* **47**, 561–571.
Polikarpov, G. G. 1966 *Radioecology of aquatic organisms*. Amsterdam: North Holland.
Polikarpov, G. G. & Ivanov, V. N. 1961 *Voprosy Ikhtiologii* **1**, 583–589. (U.K.A.E.A. translation TRG. Information series 166 [W]).
Polikarpov, G. G. & Ivanov, V. N. 1962 *Radiobiology* **2**, 207–210. (A.E.C.-tr-5429, 50–56.)
Preston, A. & Jefferies, D. F. 1967 *Hlth Phys.* **13**, 477–485.
Preston, A. & Jefferies, D. F. 1969*a* *Hlth Phys.* **16**, 33–46.
Preston, A. & Jefferies, D. F. 1969*b* *Environmental contamination by radioactive materials*, pp. 183–209. Vienna: I.A.E.A.
Price, T. J. 1965 Annual report of the Bureau of Commercial Fisheries Radiobiological Laboratory, Beaufort, N. C. *U.S. Dept. Int., U.S. Fish and Wildl. Serv., Bur., Comm. Fish. Circ.* no. 204, 10–14.
Purdom, C. E. 1966 *Disposal of radioactive wastes into seas, oceans and surface waters*, pp. 861–867. Vienna: I.A.E.A.
Rees, G. H. 1962 *Chesapeake Sci.* **3**, 29–34.
Revelle, R. & Schaefer, M. B. 1957 *The effects of atomic radiation on oceanography and fisheries*, pp. 1–25. Publication 551 National Academy of Sciences—National Research Council, Washington, D.C.
Schröder, J. H. 1969 *Mutation Res.* **7**, 75–90.
Templeton, W. L. 1966 *Disposal of radioactive wastes into seas, oceans and surface waters*, pp. 847–860. Vienna: I.A.E.A.
Templeton, W. L., Nakatani, R. E. & Held, E. 1970 Radiation effects. Section VIII of *Radioactivity in the marine environment*. National Academy of Science—National Research Council, Washington, D.C. (to be published).
Ward, E., Beach, S. A. & Dyson, E. D. 1970 *UKAEA PG Report 1013 (W)*, London: H.M.S.O.
Welander, A. D., Donaldson, L. R., Foster, R. F., Bonham, K., Seymour, A. H. & Lowman, F. G. 1949 *USAEC Doc. no. UWFL*-17.
White, J. C. 1969 Progress report of the Bureau of Commercial Fisheries Radiobiological Laboratory, Beaufort, N.C. *U.S. Dept. Int. U.S. Fish and Wildl. Serv., Bur. Comm. Fish. Circ.* no. 309, 41–47.
White, J. C. & Angelovic, J. W. 1966 *Chesapeake Sci.* **7**, 36–39.
Woodhead, D. S. 1970 *Radiat. Res.* **43**, 582–597.

Radioactivity in California Waters,[1]

January-December 1969

Bureau of Radiological Health, State of California Department of Public Health

Gross beta radioactivity in California domestic waters is monitored by the Sanitation and Radiation Laboratory of the State Department of Public Health. The importance of this program in the State's environmental surveillance activities stems from the fact that most of California's domestic water supplies are of surface origin.

[1] Data from July and October 1969 issues of Radiological Health News, State of California, Department of Public Health, Bureau of Radiological Health, 2151 Berkeley Way, Berkeley, Calif. 94704.

Radioactivity in such water supplies consists of the natural radioactivity in surface streams, radioactivity added by the discharge of sewage or by industrial waste effluents, and radioactivity from fallout, particularly fallout into open terminal or distribution reservoirs. Present efforts consist of sampling raw and treated surface waters and well waters (table 1). It should be noted that except for large metropolitan water supplies, raw water sampling is being phased out and treated water sampling being substituted or continued. This procedural change is predicated upon sampling water at the point of consumption.

Most of the supplies sampled have raw surface waters (figure 1) as their source, although a few wells, along with some water supplies that use infiltration galleries, are also sampled.

Monitoring of domestic water supplies is on a continuing basis, since it has not been possible to forecast levels of radioactivity in these supplies based upon levels in rain, snow, or surface streams. Under the present sampling schedule, monthly 500 ml samples are collected, and the total solids analyzed for alpha and beta radioactivity. In addition, 3-liter samples are collected monthly for approximately 6 months and composited for specific radionuclide analysis on a semiannual basis.

Analytical procedures

Radionuclide analyses of water are carried out in the State's Sanitation and Radiation Laboratory. Measurements of alpha and alpha-plus-beta radioactivities are made with a low-background windowless gas-flow proportional counter. Counting methods used follow those recommended by the U.S. Public Health Service (*1*).

Individual samples are evaporated to dryness and the residue ashed at 450°C. The ashed sample is dissolved and transferred to an aluminum planchet for beta-particle counting. Radioisotopic anal-

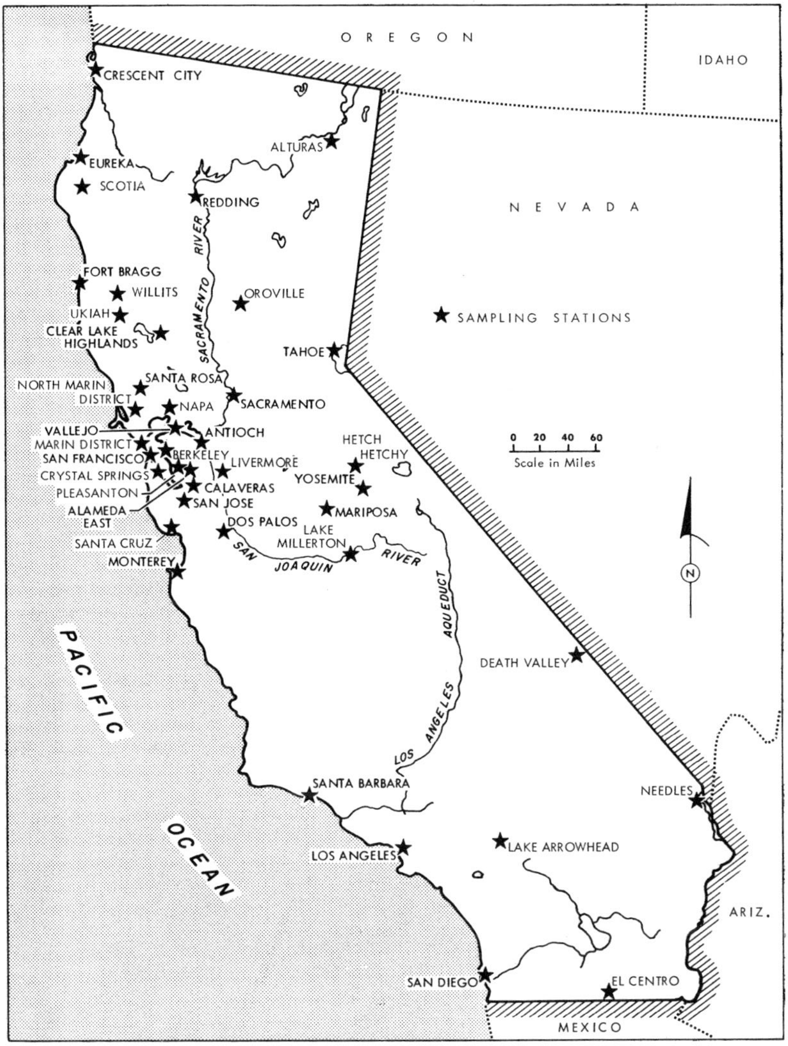

Figure 1. California surface water sampling stations

yses are determined semiannually on composite samples. Gamma-ray emitting nuclides are determined by gamma-ray spectroscopy and radium and radiostrontium by chemical separation and counting.

Sewage samples

Sewage samples are collected from many treatment plants throughout the State. Analyses of sewage effluent and sludge samples provide a means of monitoring to insure that industrial radioactive wastes discharged into sewage systems do not become excessive. The surveillance of sewage assumes greater importance as isotope licensees become more numerous and as the quantity per user increases. Analysis procedures for sewage samples are the same as the method used for domestic water. Gross radioactivity in sewage samples for January–December 1969 is presented in table 2.

The relationship between domestic water and sewage is of significance. One would expect a one-to-one radiological relationship between them unless there was some disposal of radioactive material into the sewage. The sewage to domestic water ratio has been calculated for those places where both samples are collected and analyzed. The data are in table 3. In addition to the ratio, the difference in the concentration of radioactive material in sewage and water was determined.

Looking at either the sewage-water ratio or the difference, it is apparent that some sewages in the State contain no additional radioactivity in the sewage as compared to its domestic water, and some sewages do contain additional radioactive material.

Tables 4 and 5 show a comparison of gross beta yearly average concentrations in sewage effluent and sludge for 1964 through 1969.

Discussion

Table 1 shows the monthly average beta radioactivity in the suspended-plus-dissolved solids in surface water supplies in California for January–December 1969. Following treatment, these waters are used for industrial and domestic purposes. Because alpha radioactivity in water has, in general, been undetectable or very slight, alpha radioactivity analyses are not presented. No increase in radioactivity level of surface water has been observed.

REFERENCE

(*1*) PUBLIC HEALTH SERVICE, DIVISION OF RADIOLOGICAL HEALTH. Radionuclide analyses of environmental samples, R 59–6. Radiological Health Research Activities, Robert A. Taft Sanitary Engineering Center, Cincinnati, Ohio 45226 (November 15, 1959).

Table 1. Gross beta radioactivity in California domestic waters, January–December 1969

Sampling station	Quality	Concentration[a] (pCi/liter)											
		Jan	Feb	Mar	Apr	May	June	July	Aug	Sept	Oct	Nov	Dec
Alturas	Well	31	13	8	0	0	10	8	5	17	15	NS	20
Antioch	Treated	19	17	0	8	0	0	NS	NS	2	10	NS	NS
Lake Arrowhead	Treated	25	7	10	NS	NS	NS	0	NS	NS	6	13	17
Berkeley	Treated	7	7	0	7	8	5	8	NS	7	0	2	22
Clearlake Highlands	Treated	1	28	3	12	10	9	NS	0	1	26	0	20
Crescent City	Well	0	0	1	7	0	3	3	0	NS	0	NS	2
Death Valley	Treated	32	1	16	16	4	NS	NS	NS	NS	NS	NS	35
Dos Palos	Treated	0	0	9	13	0	19	0	0	8	8	0	1
El Centro	Treated	NS	NS	NS	NS	NS	NS	NS	NS	11	18	NS	NS
Eureka	Raw	0	0	1	4	0	6	5	8	0	1	0	0
	Treated	5	42	14	0	8	18	NS	NS	NS	NS	NS	NS
Livermore	Well	NS	NS	NS	NS	6	0	4	NS	7	0	0	10
Los Angeles	Raw	NS	NS	19	NS	1	0	0	30	20	3	21	6
Marin Municipal Water District	Treated	13	35	3	2	0	5	5	3	6	9	0	0
Metropolitan Water District of Southern Calif.	Treated	28	28	0	25	18	19	51	96	36	35	30	21
Lake Millerton	Raw	7	2	0	14	0	7	5	NS	5	10	6	NS
Monterey	Treated	14	0	7	0	7	1	NS	2	NS	5	20	10
Napa	Treated	11	6	NS	NS	0	8	10	NS	13	12	0	13
Needles	Well	NS	NS	14	NS	NS	2	8	5	10	NS	NS	NS
North Marin Water District	Treated	6	NS	NS	9	2	19	NS	NS	NS	NS	NS	NS
	Raw	11	0	7	NS	NS	NS	13	NS	0	6	19	13
	Sludge[b]	16	NS	NS	NS	NS	NS	NS	NS	NS	NS	NS	NS

Oroville	Treated	0	12	4	10	3	5	NS	NS	0	NS	NS	NS
Wyandotte Irrigation	Raw	NS	NS	NS	NS	NS	NS	0	NS	14	4	NS	NS
District	Treated	NS	NS	NS	NS	NS	NS	NS	NS	0	NS	23	12
Pleasanton	Well	0	7	15	3	2	8	NS	NS	NS	NS	NS	NS
	Treated	NS	NS	NS	NS	NS	NS	5	3	4	0	0	16
Redding	Treated	17	7	6	11	10	44	0	4	3	0	0	13
Sacramento:													
American River	Treated	0	8	NS	11	0	NS	NS	NS	NS	NS	NS	NS
Sacramento	Treated	8	1	NS	11	4	NS	8	NS	0	17	13	NS
San Diego	Raw	0	0	7	24	23	3	8	NS	NS	NS	NS	NS
	Treated	23	10	NS	18	14	23	0	NS	NS	NS	NS	NS
San Francisco	Raw	6	4	80	10	10	0	NS	9	11	11	0	14
San Jose	Raw	0	NS	14	0	7	9	0	8	0	2	14	NS
Santa Barbara	Treated	0	9	20	12	6	7	6	2	14	14	0	0
Santa Cruz	Raw	0	0	0	0	0	5	0	0	8	1	0	0
	Treated	NS	NS	NS	NS	NS	NS	11	15	14	3	11	8
Santa Rosa	Well	NS	NS	NS	2	11	6	6	3	3	0	0	0
Scotia	Treated	6	11	22	6	7	20	NS	9	0	5	NS	NS
Tahoe City	Raw	NS	7	9	3	NS	NS	0	NS	NS	NS	NS	NS
Ukiah	Well	0	7	0	5	0	NS	8	7	11	1	23	NS
Vallejo	Treated	15	0	1	7	0	0	18	15	31	0	10	23
Willits	Treated	15	32	5	0	0	5	4	11	7	2	3	8
Yosemite	Treated	29	0	19	16	0	3	1	11	20	16	NS	9
Maximum		32	42	80	25	23	44	51	96	36	35	30	35
Minimum		0	0	0	0	0	0	0	0	0	0	0	0

[a] Levels below the practical reporting level (15 pCi/liter) should be regarded as statistically insignificant.
[b] Sludge reported in pCi/g (dry weight).
NS, no sample.

Table 2. Gross radioactivity in California sewage samples, January–December 1969

Sewage treatment plant	Concentration (pCi/liter)											
	Jan	Feb	Mar	Apr	May	June	July	Aug	Sept	Oct	Nov	Dec
Antioch:												
Effluent:												
Alpha	0	1	6	0	0	0	NS	NS	NS	1	NS	0
Beta	0	0	36	29	0	0	NS	NS	NS	4	NS	0
Sludge:												
Alpha	0	20	0	1	9	10	NS	NS	NS	5	7	2
Beta	24	0	15	41	0	0	NS	NS	NS	9	6	15
Bakersfield:												
Effluent:												
Alpha	3	1	0	0	0	0	0	1	1	0	0	0
Beta	0	0	18	35	5	0	2	26	15	0	21	1
Sludge:												
Alpha	NS	NS	NS	NS	NS	NS	NS	NS	NS	NS	NS	NS
Beta	NS	NS	NS	NS	NS	NS	NS	NS	NS	NS	NS	NS
El Centro:												
Effluent:												
Alpha	13	0	22	0	6	5	NS	0	18	0	2	0
Beta	28	9	1	10	8	0	NS	10	0	28	2	2
Sludge:												
Alpha	41	20	17	10	10	26	27	17	29	6	20	29
Beta	27	91	27	14	9	23	31	43	33	22	29	14
Eureka[a]												
Effluent:												
Alpha	1	5	NS	0	NS	NS	13	0	4	NS	NS	NS
Beta	0	0	NS	20	NS	NS	57	19	0	NS	NS	NS
Sludge:												
Alpha	14	3	0	7	NS	NS	3	0	2	NS	NS	NS
Beta	0	67	0	27	NS	NS	0	7	1	NS	NS	NS
Fresno:												
Effluent:												
Alpha	2	4	0	5	1	2	2	1	1	0	1	0
Beta	0	20	2	0	6	0	0	48	10	61	40	27
Sludge:												
Alpha	8	18	25	11	0	13	13	12	4	13	7	14
Beta	0	15	13	25	5	39	52	13	27	2	26	29

Livermore:												
Effluent:												
Alpha	0	0	0	6	2	0	0	0	0	0	2	3
Beta	4	0	19	9	6	21	21	43	70	20	37	40
Sludge:												
Alpha	6	12	14	6	21	8	8	11	0	17	13	44
Beta	141	52	23	102	417	235	243	218	539	16	185	401
Los Angeles City:												
Hyperion:												
Effluent:												
Alpha	5	16	22	9	13	11	18	0	0	5	2	4
Beta	0	10	33	18	12	20	50	31	20	1	8	13
Sludge:												
Alpha	9	8	15	10	11	15	12	3	12	0	5	10
Beta	14	11	10	7	19	0	4	19	19	25	31	28
Los Angeles County:												
Sanitary District:												
Effluent:												
Alpha	0	0	0	9	0	0	0	0	0	NS	4	13
Beta	0	0	28	17	18	0	0	26	28	NS	58	26
Sludge:												
Alpha	10	12	11	4	3	6	7	NS	2	9	18	5
Beta	6	27	11	13	1	0	22	NS	18	23	17	0
Oakland-East Bay mud:												
Effluent:												
Alpha	NS	0	NS	NS	0	NS	NS	0	0	0	0	5
Beta	NS	0	NS	NS	52	NS	NS	37	7	0	8	41
Sludge:												
Alpha	NS	17	NS	NS	12	NS	NS	0	8	NS	13	NS
Beta	NS	16	NS	NS	10	NS	NS	21	16	NS	18	NS
Orange County:												
Plant 1:												
Effluent:												
Alpha	2	0	0	11	17	3	0	0	NS	NS	NS	NS
Beta	0	54	55	17	0	20	36	42	NS	NS	NS	NS
Sludge:												
Alpha	28	18	5	21	15	4	7	0	NS	NS	NS	NS
Beta	20	0	18	0	23	0	1	25	NS	NS	NS	NS
Orange County:												
Plant 2:												
Effluent:												
Alpha	11	0	7	4	36	3	7	2	NS	NS	NS	NS
Beta	8	29	0	30	1	1	35	16	NS	NS	NS	NS

See footnotes at end of table.

Table 2. Gross radioactivity in California sewage samples, January–December 1969—Continued

Sewage treatment plant	Concentration (pCi/liter)											
	Jan	Feb	Mar	Apr	May	June	July	Aug	Sept	Oct	Nov	Dec
Orange County:												
Plant 2: (Continued)												
Sludge:												
Alpha	26	12	5	39	18	12	9	18	NS	NS	NS	NS
Beta	0	24	18	0	35	9	3	56	NS	NS	NS	NS
Pleasanton:[a]												
Effluent:												
Alpha	5	20	7	2	2	5	7	4	6	14	5	0
Beta	29	24	55	23	14	19	45	69	37	98	62	133
Sludge:												
Alpha	109	102	71	85	80	72	67	49	47	33	56	53
Beta	171	113	121	192	313	217	160	222	354	363	418	276
Sacramento:												
Effluent:												
Alpha	0	0	16	2	1	0	4	1	NS	0	8	1
Beta	0	0	36	3	16	0	0	22	NS	1	2	12
Sludge:												
Alpha	29	17	3	7	7	12	8	5	NS	11	4	13
Beta	8	15	17	15	3	10	20	28	NS	19	14	23
Salinas:												
Effluent:												
Alpha	0	4	48	10	12	0	0	0	0	8	17	2
Beta	15	0	0	4	0	41	41	53	46	13	41	18
Sludge:												
Alpha	19	61	10	8	17	7	8	13	15	6	8	9
Beta	16	0	38	0	13	16	24	19	72	25	47	30
San Bernardino:												
Effluent:												
Alpha	NS	NS	NS	NS	NS	NS	NS	NS	NS	NS	NS	NS
Beta	NS	NS	NS	NS	NS	NS	NS	NS	NS	NS	NS	NS
Sludge:												
Alpha	NS	NS	NS	NS	NS	NS	NS	NS	NS	NS	NS	NS
Beta	NS	NS	NS	NS	NS	NS	NS	NS	NS	NS	NS	NS

San Diego:												
Effluent:												
Alpha	13	10	1	1	2	13	0	1	5	NS	1	15
Beta	0	13	42	23	4	21	36	21	84	NS	34	44
Sludge:												
Alpha	NS	6	0	15	22	14	14	14	16	NS	30	17
Beta	NS	11	3	30	42	13	26	35	20	NS	40	7
San Francisco:												
North Point:												
Effluent:												
Alpha	12	0	0	0	7	2	0	0	35	0	0	0
Beta	13	27	0	22	27	8	23	10	46	34	67	12
Sludge:												
Alpha	NS	NS	NS	NS	NS	NS	NS	NS	NS	NS	NS	NS
Beta	NS	NS	NS	NS	NS	NS	NS	NS	NS	NS	NS	NS
Richmond Sunset:												
Effluent:												
Alpha	0	6	0	0	0	0	0	0	0	0	0	0
Beta	0	0	26	10	55	0	0	9	17	8	11	0
Sludge:												
Alpha	13	5	4	6	4	9	9	21	0	NS	9	11
Beta	0	34	15	15	10	0	4	35	23	NS	39	13
Southeast:												
Effluent:												
Alpha	0	12	19	0	0	0	0	0	0	1	0	19
Beta	23	38	26	24	19	25	40	0	0	24	0	42
Sludge:												
Alpha	19	15	3	7	0	19	NS	14	13	NS	3	NS
Beta	15	34	25	18	26	0	NS	15	18	NS	8	NS
San Jose:												
Effluent:												
Alpha	NS	42	NS	1	14	8	NS	0	0	NS	2	NS
Beta	NS	1	NS	1	20	5	NS	5	11	NS	18	NS
Sludge:												
Alpha	NS	12	NS	16	10	59	59	24	9	NS	NS	13
Beta	NS	84	NS	154	215	412	478	223	8	NS	NS	25
Vallejo:												
Effluent:												
Alpha	0	11	2	6	1	0	1	1	0	0	10	0
Beta	20	52	50	23	12	8	26	6	38	12	14	7
Sludge:												
Alpha	6	7	6	13	3	6	22	2	4	3	11	3
Beta	9	8	9	0	9	10	6	0	20	0	24	0

[a] Average of Hill, McCullen, and Murray Street treatment plants.
[b] Average of triplicate samples.
NS, no sample.

Table 3. Comparison of gross beta radioactivity in domestic water and sewage effluent in California, 1969

Sampling station	Average concentration (pCi/liter)			Sewage to water ratio
	Domestic water	Sewage effluent	Difference	
Antioch	7	9	2	1.3
East Bay mud	7	20	13	2.9
Eureka	10	16	6	1.6
Livermore	4	24	20	6.0
Los Angeles City	13	18	5	1.4
Pleasanton	5	51	46	10.2
Sacramento	6	8	2	1.3
San Diego	9	29	20	3.2
San Francisco:				
North Point	12	24	12	2.0
Richmond Sunset	12	12	0	1.0
South East	12	22	10	1.8
San Jose	7	9	2	1.3
Vallejo	7	22	15	3.1

Table 4. Gross beta radioactivity in California sewage effluent, 1964–1969

Treatment plant	Average concentration (pCi/liter)						6-year-average
	1964	1965	1966	1967	1969	1969	
Antioch	35	15	8	14	3	9	14
Bakersfield	47	41	2	18	9	10	21
El Centro	18	11	15	29	24	9	18
Eureka	89	23	9	7	9	16	26
Fresno	15	25	7	8	12	18	14
Livermore	22	35	29	24	32	24	28
Los Angeles City:							
Hyperion	18	NS	18	20	12	18	18
Los Angeles County	20	29	30	38	19	18	31
East Bay mud	29	28	27	14	27	20	24
Orange County:							
Plant 1	33	25	19	37	21	28	27
Plant 2	95	30	22	36	25	15	37
Pleasanton	124	34	33	29	21	51	49
Sacramento	19	7	8	21	22	8	14
Salinas	16	12	20	NS	11	23	18
San Diego	36	26	48	36	NS	29	37
San Francisco:							
North Point	45	11	18	25	12	24	23
Richmond Sunset	28	15	20	9	11	12	16
Southeast	67	70	38	46	28	22	45
San Jose	26	23	17	12	31	9	20
Vallejo	26	17	16	12	13	22	18

NS, no sample.

Table 5. Gross beta radioactivity in sludge in California sewage treatment plants, 1964–1969

Treatment plant	Average concentration (pCi/dry g)						6-year-average
	1964	1965	1966	1967	1968	1969	
Antioch	20	12	10	44	8	14	18
El Centro	37	20	15	21	21	30	24
Eureka	28	22	8	1	7	15	14
Fresno	52	15	15	28	17	21	25
Livermore	76	64	54	95	85	214	98
Los Angeles City:							
Hyperion	9	NS	26	22	14	16	17
Los Angeles County	15	15	23	13	7	13	14
East Bay mud	46	20	16	14	8	16	20
Orange County:							
Plant 1	56	29	20	22	5	11	24
Plant 2	35	29	15	31	19	20	25
Pleasanton	460	400	576	320	210	244	370
Sacramento	30	14	14	19	12	15	17
Salinas	30	42	21	NS	16	25	27
San Diego	54	27	18	19	24	21	27
San Francisco:							
Richmond Sunset	62	65	24	26	18	17	35
Southeast	46	30	34	36	28	18	32
San Jose	40	14	25	91	164	200	89
Vallejo	55	23	13	13	10	8	20

NS, no sample.

Radioactivity in New York Surface Water July-December 1969 and January-June 1970

Bureau of Radiological Pollution Control
New York State Department of Environmental Conservation

In 1955, the New York State Department of Health began a program to determine the amount of radioactivity in water used for public consumption. On July 1, 1971 this program was transferred to the newly formed New York State Department of Environmental Conservation. Radioactivity in water may arise from any one or a combination of the following sources: the natural mineral content of water (background), atmospheric fallout, or nuclear industry operations.

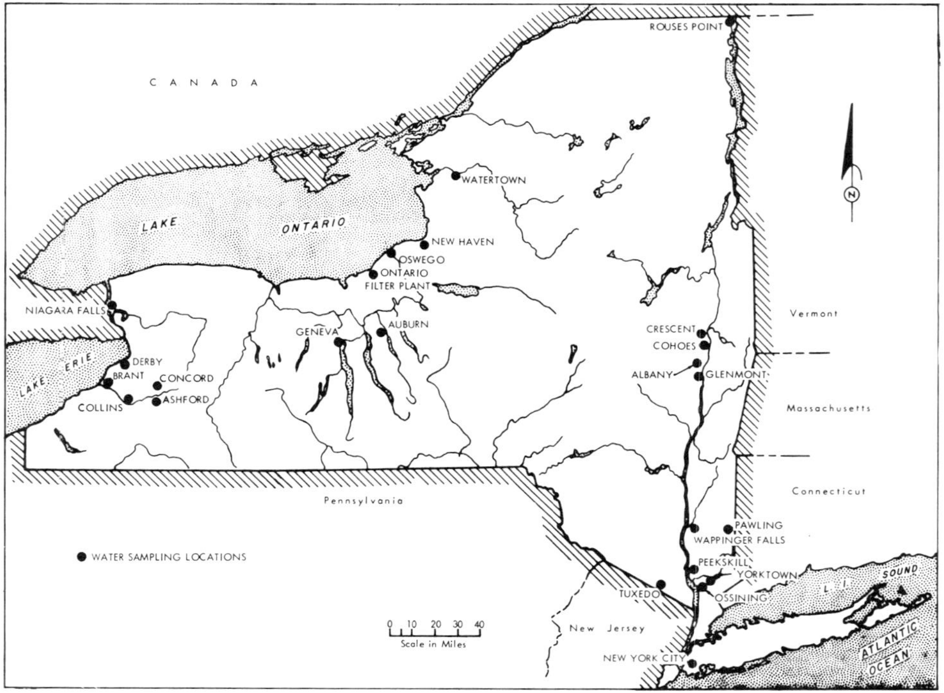

Figure 1. New York water sampling locations

Analytical procedures

A measured quantity of water, usually 500 ml, is evaporated and the residue is analyzed for its gross beta component in an end-window, gas-flow proportional counter.

Strontium and alkaline earths are precipitated as carbonates from a 500-ml sample. Iron and rare earths are removed by hydroxide scavenging, while barium is precipitated as a chromate. Strontium is finally precipitated as a sulphate from a pH controlled EDTA solution. Calcium and yttrium remain in solution as EDTA complexes (*1–2*).

Strontium-90 is determined by yttrium-90 ingrowth counting of the final precipitate at less than 6 hours after precipitation and again at greater than 50 hours using a low background (less than 1 cpm) gas-flow, proportional beta-particle counter. Strontium-89 is estimated by taking the difference between the total strontium and the strontium-90 radioactivity.

Chemical recovery is between 70 and 75 percent and results in a minimum detectable radioactivity of 3 pCi/liter ± 100 percent at the 95-percent confidence level.

Tritium in water is determined by liquid scintillation counting of 3 ml of distilled sample in 17 ml of scintillator solution. The scintillator solution consists of 4 g PPO, 0.25 g naphthalene dissolved in 1 liter of dioxane. The sample is counted twice for a total of 100 minutes using either a Packard Tri-Carb model 3315 or a Beckman LS 200B liquid scintillation spectrometer. Minimum sensitivities at the 95-percent confidence level ± 100 percent are 1,000 pCi/liter for the Tri-Carb and 500 pCi/liter for the Beckman.

Discussions and results

Radioactivity levels in water for July-December 1969 continued to remain low throughout the State with the exception of Buttermilk and Cattaraugus Creeks downstream from Nuclear

Table 1. Gross beta radioactivity in New York raw surface water, July–December 1969[a]

Location	Gross beta radioactivity (pCi/liter)							
	July–September 1969				October–December 1969			
	Number of samples	Average	Maximum	Minimum	Number of samples	Average	Maximum	Minimum
Albany	3	2	3	2	2	3	3	2
Ashford (Buttermilk Creek at Fox Valley Road)	3	3	5	ND	3	4	5	4
(Buttermilk Creek at Thomas Corners)[b]	14	2,587	5,446	173	13	2,135	5,060	379
(Cattaraugus Creek at Bigelow Bridge)	3	2	3	ND	3	4	6	3
(Cattaraugus Creek at Felton Bridge)	10	342	915	132	8	322	593	206
Brant (Cattaraugus Creek)	14	136	235	85	11	121	225	39
Collins (Cattaraugus Creek)	14	179	419	82	11	168	281	109
Concord (Cattaraugus Creek)	58	279	506	143	12	351	535	167
(Cattaraugus Creek—Site 042)	NS	NS	NS	NS	55	339	926	71
Crescent (Crescent Dam)	10	4	5	3	11	3	5	2
Geneva (Seneca Lake)	2	4	4	4	3	5	8	4
Glenmont (Hudson River)	8	4	5	3	10	4	10	3
New Haven (Lake Ontario)	3	4	5	4	3	4	4	3
New York City	NA	NA	NA	NA	5	2	4	ND
Niagara Falls (West Branch, Niagara River)	2	4	4	4	3	4	5	3
Ontario Filter Plant	3	5	6	4	2	4	4	3
Ossining (Indian Brook Reservoir)	3	4	7	ND	3	4	7	ND
(Sing Sing) (Hudson River)	12	34	60	12	11	35	71	14
Oswego	3	4	5	3	2	4	4	3
Pawling (Pond at United Nuclear)	4	4	5	3	5	3	4	2
Peekskill (Camp Field water supply)	3	ND	ND	ND	3	4	13	ND
(Hudson River)	12	13	30	ND	11	14	32	ND
Tuxedo (Indian Kill)	3	7	11	5	3	4	6	3
Watertown (Black River)	4	4	5	3	3	3	3	3
Yorktown (Croton Reservoir)	3	4	4	3	3	4	5	3

[a] Excluding tritium.
[b] This station is on the Nuclear Fuels Services reprocessing plant site.
NA, no analysis.
ND, nondetectable.
NS, no sample.

Fuel Services. The gross beta and strontium-90 levels in Cattaraugus Creek for October–December were higher than for the July–September 1969 period. The average concentration of strontium-90 in Cattaraugus Creek at Springville Dam was 98 pCi/liter for the October–December 1969 period as compared to 25 pCi/liter for the July–September 1969 period. A new water sampling station was added, the Ontario filter plant in Wayne County (figure 1).

During July–September 1969, some water samples were collected from Cattaraugus Creek at Felton Bridge, site 032, instead of the Springville Dam, site 042, due to the annual cleaning operation to flush out silt from behind the dam. A grab sample upsteam from the dam at Felton Bridge was collected on September 17, 1969 and contained 915 pCi/liter gross beta radioactivity. This is higher than the 600 pCi/liter considered to be the allowable limit for gross beta radioactivity.

Beta and gamma radiation from the isotope ruthenium-106 has started to appear in significant concentrations in the water courses below Nuclear Fuel Services resulting in increased gross beta levels. This increase apparently results from the processing of spent fuel that had a shorter storage period than fuels previously processed. Ruthenium-106 has a good fission yield and will decay in spent fuel with a half-life of 1 year, so that the longer spent fuel is stored before processing, the less the ruthenium-106 activity. The amount of ruthenium-106 in spent fuel stored for 5 years will be lower by a factor of 16 than spent fuel stored for 1 year. The maximum concentration allowed for ruthenium-106 in drinking water is 10.0 nCi/liter and the most important human exposure is the exposure to the lower large intestine. Ruthenium-106 does not concentrate in any particular organ of the human body nor has it been found to significantly reconcentrate in edible fish, milk, or other food pathways for human exposure.

The number of samples and the quarterly average, maximum, and minimum gross beta-

Table 2. Strontium-90 concentration in New York surface waters, July–December 1969

Location	Radionuclide concentration (pCi/liter)							
	July–September 1969				October–December 1970			
	Number of samples	Average	Maximum	Minimum	Number of samples	Average	Maximum	Minimum
Ashford (Buttermilk Creek at Thomas Corners)[a]	14	308	1,111	80	13	703	1,658	182
(Cattaraugus Creek at Felton Bridge)	3	135	194	97	1	72	—	—
(Cattaraugus Creek at Bigelow Bridge)	3	ND	ND	ND	3	ND	ND	ND
(Buttermilk Creek at Fox Valley Road)	3	ND	ND	ND	3	ND	ND	ND
Brant (Cattaraugus Creek)	11	14	32	ND	12	36	62	21
Collins (Cattaraugus Creek)	13	14	27	ND	12	43	75	24
Concord (Springville Power Dam on Cattaraugus Creek)	12	25	55	10	12	98	208	51
(Springville Dam—Site 042)[b]	NS	NS	NS	NS	5	114	224	53
Hudson River at Standard Brands	10	ND	ND	ND	12	ND	ND	ND
Hudson River at Sing Sing	10	ND	ND	ND	12	ND	ND	ND

[a] This station is in the Nuclear Fuels Services reprocessing plant site.
[b] Daily continuous sample taken.
ND, nondetectable.
NS, no sample.

particle concentrations in New York surface water for July–December 1969 are given in table 1. Strontium-90 in surface water is shown in table 2.

Radioactivity levels in water for the January–June 1970 period continue to remain low throughout the State with the exception of Buttermilk and Cattaraugus Creeks, downstream of Nuclear Fuel Services, Incorporated. Two daily samples from Site 042 (Cattaraugus Creek at Springville Dam) were over 600 pCi/liter which is considered to be the allowable AEC limit for gross beta radioactivity.

The water sample collected on May 26, 1970 from Cattaraugus Creek downstream at Gowanda contained a gross beta concentration of 1,166 pCi/liter. During this period, the stream beds in Cattaraugus and Buttermilk Creeks were disturbed by bulldozer operations which probably resuspended clay and silts which flowed downstream. Both suspended and dissolved solids portions of the sample were analyzed for radioactivity and most of the gross radioactivity was found in the suspended solids portion of the sample.

Table 3. Radioactivity in an algae sample collected July 9, 1970, Croton Reservoir

Radionuclide	Concentration (pCi/kg wet weight)
Ruthenium-rhodium-106	2,916
Cesium-137	479
Zirconium-niobium-95	1,484
Cobalt-60	nondetectable

Table 4. Gross beta radioactivity in water samples Croton Reservoir

Date (1970)	Concentration (pCi/liter)
April 16	5
May 18	80
June 16	7
July 2	4
July 15	4

Table 5. Gross beta radioactivity in New York raw surface water, January–June 1970[a]

Location	Gross beta radioactivity (pCi/liter)							
	January–March 1970				April–June 1970			
	Number of samples	Average	Maximum	Minimum	Number of samples	Average	Maximum	Minimum
Albany	3	2	2	2	3	2	3	2
Ashford (Buttermilk Creek at Fox Valley Road)[a]	3	13	31	2	3	4	5	3
(Buttermilk Creek at Thomas Corners)[b]	13	1,940	3,576	530	13	1,183	2,564	156
(Cattaraugus Creek at Bigelow Bridge)	3	8	19	2	3	4	7	3
Brant (Cattaraugus Creek)	7	88	125	55	13	68	139	22
Collins (Cattaraugus Creek)	13	108	161	33	13	154	1,166	29
Concord (Cattaraugus Creek)	13	247	367	142	13	155	246	61
Geneva (Seneca Lake)	2	4	4	3	3	4	5	4
Glenmont (Hudson River)	13	4	7	2	13	4	5	3
New Haven (Lake Ontario)	4	4	4	3	2	4	5	4
New York City	7	3	3	2	7	4	5	2
Niagara Falls (West Branch Niagara River)	2	4	4	4	3	3	4	3
Ossining (Indian Brook Reservoir)	3	4	5	4	3	7	13	3
(Sing Sing) (Hudson River)	16	21	48	ND	13	17	36	4
Oswego	3	4	4	3	3	4	4	4
Pawling (Pond at United Nuclear)	3	2	2	2	5	4	5	3
Peekskill (Camp Field Water Supply)	3	4	9	ND	3	6	10	4
(Hudson River)	14	7	19	ND	13	9	20	2
Tuxedo (Indian Kill)	3	5	5	4	3	5	6	5
Watertown (Black River)	3	3	4	3	3	4	5	3
Yorktown (Croton Reservoir)	3	5	6	4	3	31	80	5

[a] Excluding tritium.
[b] This station is on the Nuclear Fuels Services reprocessing plant site.
ND, nondetectable.

Table 6. Strontium-90 concentration in New York surface waters, January–June 1970

Location	Radionuclide concentration (pCi/liter)							
	January–March 1970				April–June 1970			
	Number of samples	Average	Maximum	Minimum	Number of samples	Average	Maximum	Minimum
Ashford (Buttermilk Creek at Thomas Corners)[a]	11	800	1,624	311	14	459	1,351	88
(Cattaraugus Creek at Bigelow Bridge)	3	ND	ND	ND	3	ND	ND	ND
(Buttermilk Creek at Fox Valley Road)	3	ND	ND	ND	3	ND	ND	ND
Brant (Cattaraugus Creek at Irving)	7	39	85	19	14	25	49	ND
Collins (Cattaraugus Creek at Gowanda)	11	47	94	18	14	30	72	8
Concord (Springville Power Dam on Cattaraugus Creek)	12	89	148	54	13	48	97	ND
(Springville Dam—Site 042)[b]	2	114	139	89	2	62	92	32
Derby (Sturgeon Point Station)	1	ND	ND	ND	3	ND	4	ND
Peekskill (Hudson River at Standard Brands)	12	ND	ND	ND	14	ND	ND	ND
(Hudson River—Sing Sing)	12	ND	ND	ND	15	ND	ND	ND

[a] This station is in the Nuclear Fuels Services reprocessing plant site.
[b] Daily continuous sample taken.
ND, nondetectable.

A grab water sample collected on May 18, 1970 from Croton Reservoir at Taconic showed a gross beta concentration of 80 pCi/liter. An isotopic gamma analysis was made on this sample, and ruthenium-106 was nondetectable. Zirconium-95 was found in a concentration of 53 pCi/liter. An algae sample was collected July 9, 1970 at the same sampling point, and the results are shown in table 3. Gross beta results of grab water samples taken from this sampling point, around the period of the relatively high result, are given in table 4.

It was concluded that the water sample with the high result was collected too close to the shoreline in shallow water and some algae was included in the water sample. The radioactivity found in the water sample and in the algae sample appears to have originated from fallout associated with atmospheric weapons testing. This sampling point has been changed to deeper water in the Croton Reservoir in order to obtain a more representative water sample in the future.

Gross beta-particle concentrations for the January–June 1970 period are given in table 5; strontium-90 concentrations are given in table 6.

Tritium concentration values for July 1969–June 1970 are given in tables 7 and 8. Tritium, the radioactive isotope of hydrogen, a very low energy beta-particle emitter, is released to the water courses during the reprocessing of nuclear fuel. The tritium concentrations in Cattaraugus and Buttermilk Creeks reflected contributions to the streams from the Nuclear Fuels Services reprocessing plant.

REFERENCES

(*1*) PORTER, C., D. CAHILL, R. SCHNEIDER, P. ROBBINS, W. PERRY, and B. KAHN. Improved determination of strontium-90 in milk by ion-exchange method. Anal Chem 33:1306–1308 (September 1961).

(*2*) WELFORD, G. and D. SUTTON. Paper presented at the American Chemical Society meeting in New York City, September 1960.

Table 7. Tritium concentrations of New York surface waters, July–December 1969

Location	Concentration (nCi/liter)							
	July–September 1969				October–December 1969			
	Number of samples	Average	Maximum	Minimum	Number of samples	Average	Maximum	Minimum
Albany	3	ND	ND	ND	2	1.7	2.3	1.1
Ashford (Buttermilk Creek at Fox Valley Road)	2	ND	ND	ND	3	ND	ND	ND
(Buttermilk Creek at Thomas Corners)[a]	14	117.8	211.9	8.1	13	147.8	406.9	6.9
(Cattaraugus Creek at Bigelow Bridge)	3	ND	ND	ND	3	ND	ND	ND
(Cattaraugus Creek at Felton Bridge)	11	10.1	30.1	4.0	7	7.0	8.5	4.7
Brant (Cattaraugus Creek)	14	8.6	21.9	3.9	11	10.2	29.2	2.1
Collins (Cattaraugus Creek)	14	10.6	30.8	3.8	11	12.0	35.3	ND
Concord (Springville Power Dam on Cattaraugus Creek)	58	14.8	36.6	6.1	12	27.6	70.0	5.8
(Springville Dam-Site 042)[b]	NS	NS	NS	NS	55	26.8	98.6	2.6
Crescent (Crescent Dam)	11	ND	1.3	ND	10	ND	2.8	ND

[a] This station is on the Nuclear Fuels Services reprocessing plant site.
[b] Daily continuous sample.
ND, nondetectable.
NS, no sample.

Table 8. Tritium concentration of New York surface waters, January–June 1970

Location	Concentration (nCi/liter)							
	January–March 1970				April–June 1970			
	Number of samples	Average	Maximum	Minimum	Number of samples	Average	Maximum	Minimum
Albany	3	ND	1.4	ND	3	ND	ND	ND
(Colonie Filtration Plant)	4	ND	2.5	ND	14	1.1	6.3	ND
Ashford (Buttermilk Creek at Fox Valley Road)	3	ND	ND	ND	3	ND	1.7	ND
(Buttermilk Creek at Thomas Corners)[a]	13	179.6	430.6	17.0	13	243.0	802.8	2.3
(Cattaraugus Creek at Bigelow Bridge)	3	ND	1.7	ND	3	ND	2.1	ND
Brant (Cattaraugus Creek)	7	10.5	16.3	2.4	13	10.6	25.7	ND
Collins (Cattaraugus Creek)	13	11.8	26.1	2.4	13	10.0	27.6	ND
Concord (Springville Power Dam on Cattaraugus Creek)	13	22.0	44.3	5.3	13	18.6	39.6	ND
(Springville Dam-Site 042)[b]	62	22.9	98.6	1.9	62	18.0	54.4	ND
Crescent (Crescent Dam)	5	ND	1.2	ND	NS	NS	NS	NS
Derby (Sturgeon Point Station)	NS	NS	NS	NS	3	ND	1.8	ND
Oswego (City Hall Tap)	NS	NS	NS	NS	3	ND	1.5	ND
New Haven (Demster Peach Road)	NS	NS	NS	NS	3	ND	ND	ND
Rhinecliff (Hudson River)	NS	NS	NS	NS	13	ND	2.9	ND
Wappinger (Hudson River)	NS	NS	NS	NS	13	ND	2.9	ND

[a] This station is on the Nuclear Fuels Services reprocessing plant site.
[b] Daily continuous sample taken.
ND, nondetectable.
NS, no sample.

Radiostrontium in Tap Water, July-December 1970

Health and Safety Laboratory
U.S. Atomic Energy Commission

The Health and Safety Laboratory has performed analyses for strontium-90 in tap water at New York City since August 1954. Samples of tap water are collected daily so that by the end of the month a composite of at least 100 liters is available for analysis. Cesium-137 determinations were begun in January 1964. The analytical methods used at the laboratory are given in the Health and Safety Laboratory Manual of Standard Procedures (*1*).

Strontium-90 concentration and cesium-137 to strontium-90 ratios in New York City tap water for July through December 1970 are presented in table 1. These results appear graphically in figure 1.

A decreasing trend has been observed in the strontium-90 concentrations since the July 1963 peak. The maximum strontium-90 concentrations observed are below the acceptable limit as set forth in the interstate carrier drinking water standards (*2*).

Table 1. Radiostrontium in New York City tap water July-December 1970

Sampling months 1970	Strontium-90[a] (pCi/liter)	$^{137}Cs/^{90}Sr$
July	0.72	0.13
August	.64	.14
September	.64	.13
October	.62	.14
November	.70	.14
December	.68	.11

[a] Approximately 100 liters per sample.

REFERENCES

(*1*) U.S. ATOMIC ENERGY COMMISSION. Manual of standard procedures, 40:E-38-01-16. Health and Safety Laboratory, U.S. Atomic Energy Commission, 376 Hudson Street, New York, N.Y. 10014.

(*2*) FEDERAL REGISTER RULES AND REGULATIONS. Title 42–Public Health, Chapter 1, Public Health Service, Department of Health, Education, and Welfare; Part 72, Interstate Quarantine, Subpart J, Drinking Water Standards 27:2154–2155. Superintendent of Documents, Government Printing Office, Washington, D.C. 20402 (March 6, 1962).

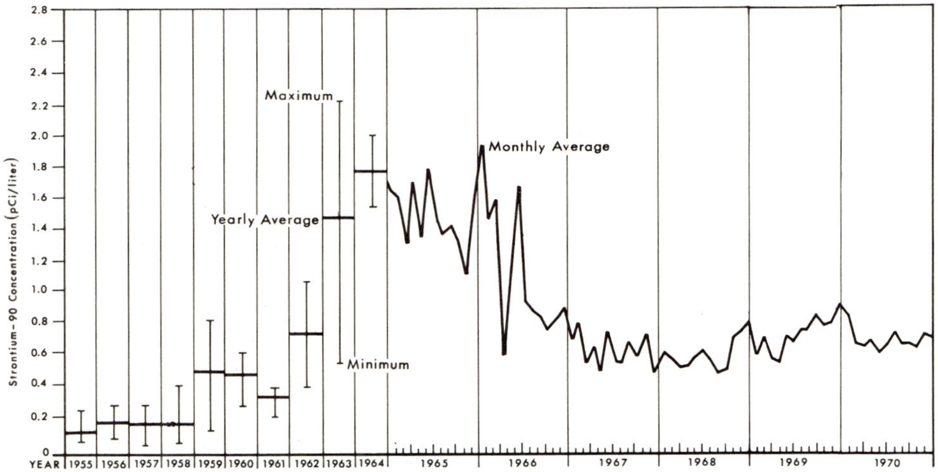

Figure 1. Strontium-90 concentrations in New York City tap water, 1955–December 1970

ON ^{134}Cs IN RAINWATER FROM 1960 TO 1969

A. A. CIGNA, G. F. CLEMENTE and F. G. GIORCELLI

INTRODUCTION

THE PURPOSE of this paper is a critical analysis of the mechanisms of production and deposition of ^{134}Cs, on the basis of all the available data.

In the group of the world-wide distributed artificial radionuclides[1] ^{134}Cs is one of the most difficult to measure owing to its very low concentration and the interference of other gamma emitters.

Furthermore, as ^{134}Cs seems to be one of the major contaminants, together with ^{60}Co, released by some nuclear plants, it is important to know the level of a world-wide contamination essentially free from the contribution of local nuclear plants.

The published data[2–6,9] on ^{134}Cs suitable for a determination of the deposition of this nuclide are rather few. Therefore this radionuclide was measured in the fallout samples collected at the Casaccia Nuclear Center (Rome, Italy) since 1960.

METHODS

The measurements were performed by means of a gamma–gamma sum coincidence spectrometer,[7] adjusted as follows:

—Coincident gamma emission of ^{134}Cs:	604; 797 keV
—Sum gamma emission:	1402 keV (window width 1260–1540 keV)
—Coincidence considered between sum gamma emission and:	797 keV (window width 720–880 keV)
Gamma emission measured by the multichannel analyzer:	604 keV (window width 540–670 keV)
—Background in the 604 keV region:	cpm (0.105 $\pm$ 0.003)†
—Counting efficiency for ^{134}Cs	

in the above reported conditions: 0.0208 counts/min/pCi.

† All the errors reported in this paper are expressed as Standard Deviation.

The original program was to perform ^{134}Cs determinations on monthly samples collected at the Casaccia Nuclear Center. Nevertheless, some preliminary measurements brought to evidence a heavy interference due to natural ^{214}Bi present in the samples. In fact, a typical result in ^{134}Cs counting conditions was:

counting rate:	0.125 ± 0.01 counts/min
background:	0.105 ± 0.003 counts/min
contribution due to ^{214}Bi:	0.021 ± 0.013 counts/min.

As can be seen, in spite of the relatively small amount of ^{214}Bi in the samples, (5–10 pCi) the interference due to this nuclide is such as to make uncertain the determination of ^{134}Cs. It must be observed that the amount of ^{214}Bi mentioned above is accounted for by the high radium content of the local soïl (volcanic rock). To overcome these difficulties, aliquots of the monthly samples were pooled together into yearly ones. The first composite sample grouped monthly samples (December 1960–September 1961) before the arrival of fresh debris produced after the first moratorium, while in the second one the remaining monthly samples of 1961 were included. A chemical separation of Cs was then performed on the composite samples by the AMP method.[8] The ^{137}Cs present in the sample, measured by gamma spectrometry, was used as a tracer for the determination of the chemical yield. The measurements of all the yearly pooled samples were performed during 1969–1970 by the coincidence apparatus quoted before and the results, corrected for physical decay, were referred to the collection period.

RESULTS AND DISCUSSION

The results obtained are reported in Table 1. Mean values, weighted according to the inverse of the variance and the length of the sampling period, were calculated both on the yearly rates of deposition and on the $^{134}Cs/^{137}Cs$ ratios reported in Table 1. The average yearly rate of deposition was 33 ± 7 $\mu Ci/km^2$ during the period September 1961–1965 while in the period 1966–1969 it decreased to 2.2 ± 1.0 $\mu Ci/km^2$.

As can be seen in Table 1, the $^{134}Cs/^{137}Cs$ ratio of the activities deposited per year is fairly constant during the whole period under consideration, the mean weighted value (computed as described before) being $(1.4 \pm 0.3) \times 10^{-3}$. In Fig. 1 are reported some values of the

*Table 1. Rate of ^{134}Cs ground deposition and comparison with ^{137}Cs deposition at the Casaccia Nuclear Center, Rome (Italy)**

Period	^{134}Cs $\mu Ci.km^{-2}.yr^{-1}$	$\frac{^{134}Cs}{^{137}Cs} \times 1000$
December 1960–September 1961	0 ± 10.0	0 ± 4.7
September 1961–December 1961	32.8 ± 23.4	4.3 ± 3.1
1962	54.6 ± 20.8	2.2 ± 0.8
1963	22.1 ± 16.0	0.7 ± 0.5
1964	31.2 ± 10.6	1.6 ± 0.5
1965	36.2 ± 19.9	3.8 ± 2.1
1966	8.6 ± 9.2	2.3 ± 2.5
1967	1.4 ± 2.6	0.9 ± 1.6
January 1968–November 1968	0.5 ± 2.3	0.2 ± 1.1
December 1968	26.8 ± 15.6	17.9 ± 10.4
1969	2.6 ± 1.2	1.9 ± 0.9

* The errors given in the Table represent the standard deviation due to the counting statistics. Other sources of error (chemical yield, counter calibration) gave a negligible contribution to the total error.

The experimental values for the periods shorter than 1 yr have been multiplied by a normalizing factor equal to $12/n$, n being the length of the period considered, expressed in months.

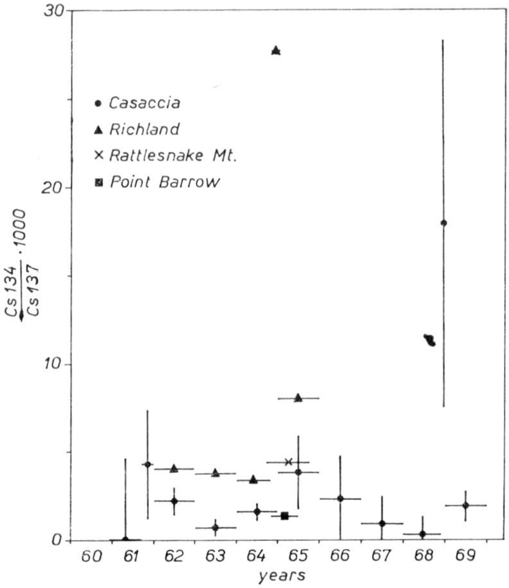

FIG. 1. $^{134}Cs/^{137}Cs$ ratios. For the errors of Italian data see note to Table 1; the errors affecting the other values are unknown. Data obtained at Richland and at Point Barrow refer to air, while the ones obtained at Casaccia and at Rattlesnake Mt. refer to rainwater.

$^{134}Cs/^{137}Cs$ ratio obtained at Casaccia (Italy); Richland, Washington;[3–5] Rattlesnake Mt., Washington[9] and Point Barrow, Alaska.[5] It can be seen that the values are fairly consistent, with very few exceptions (October–December 1964 and 1965, Richland, Washington; December 1968, Casaccia, Italy). The first of them, formerly reported as due to local contamination[5,6] could also be attributed to the contribution of the first Chinese nuclear test (16 October, 1964); the last of them is due to a release by Project Schooner test (8 December, 1968) as will be pointed out later for both cases. Furthermore, the only available value of annual ^{134}Cs deposition (Rattlesnake Mt.)[9] compared with the one obtained at the Casaccia Nuclear Center in the same year supports the hypothesis of a rather uniform deposition of ^{134}Cs in the Northern hemisphere.

The data plotted in Fig. 2 seem to indicate a definite relationship between the ^{134}Cs deposition after 1961 and the nuclear explosions of 1961 and 1962 as was pointed out by PERSSON.[1] Such a relationship is not contradicted by the data obtained before the resumption of nuclear explosions after the first moratorium. In fact, the high values[10] both of ^{134}Cs concentration in air (= 0.004 dis/min/m^3) and of the $^{134}Cs/^{137}Cs$ ratio (= 86 per mille)* measured in March 1957 at St. Louis, Missouri, show that a major injection of ^{134}Cs occurred before 1957. There is evidence that such an injection was caused by some nuclear test in the period 15 December 1955–1 May 1956.[10] No information is available on the amount of ^{134}Cs which could have been produced by the nuclear

* These values were recalculated assuming that: (gamma emissions at 0.8 MeV)/(disintegrations) = 0.97 as reported in recent literature.[11]

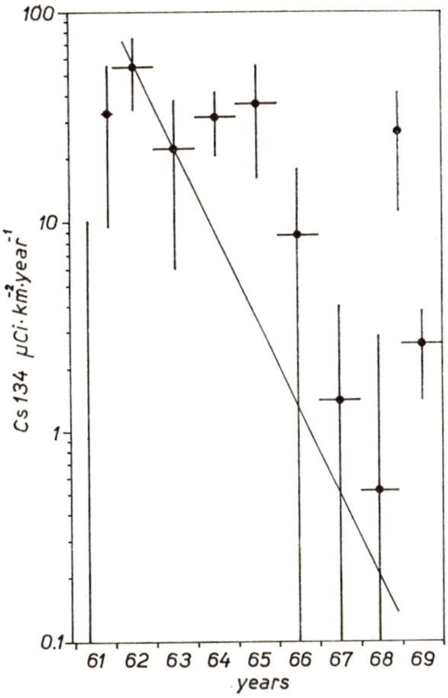

FIG. 2. ^{134}Cs rate of deposition (μCi.km^{-2} $\times$ yr^{-1}) at Casaccia Nuclear Center, Rome, Italy. A line corresponding to a mean residence half-time of 1 yr is drawn.

tests occurring in the period 1957–1958. Anyway, ^{134}Cs concentration in air, and therefore in rainwater, decreased rapidly and, immediately after 1960, it was below the detectability limits (see Fig. 2).

Following the resumption of nuclear tests in 1961, the ^{134}Cs concentration in air increased rapidly, and reached, after 1962,[3] values of about an order of magnitude smaller than the value measured in 1957. It is therefore reasonable to suppose that the explosion(s) responsible for the massive production of ^{134}Cs before 1957, had a much higher yield for this radionuclide than the explosions of the 1961–1962 series. The experimental values of the ^{134}Cs/^{137}Cs ratio in biological samples, collected at different sampling points during 1958–1963,[1,3,12–14] are rather high (decreasing from 40 to 5 per mille) while the corresponding values in air and in rainwater since 1960 are definitively smaller (12; this paper, Fig. 1). This fact could be due to the deposition and accumulation of ^{134}Cs in soil. Successively a sharp decrease in the ^{134}Cs/^{137}Cs ratio was observed[1–3,15] on account of the deposition of the ^{137}Cs produced by the 1961–1962 tests.

The experimental data of ^{134}Cs deposition (Fig. 2) could suggest the hypothesis of an injection of this nuclide during the year 1964 (first Chinese bomb: 16 October 1964). This hypothesis is supported by the sharp increase of ^{134}Cs concentration in air observed at Richland, Washington[4] at the end of 1964, together with an analogous increase of the concentrations of short-lived fission products (^{140}Ba–^{140}La, ^{103}Ru).

On account of the low yield of the first Chinese bomb, it is reasonable to suppose that its debris did not reach the stratosphere, as confirmed by high altitude measurements.[23,24] The high values of ^{134}Cs concentrations in air[5] and in rainwater (Fig. 2) measured in 1965 might be the consequence of ^{134}Cs production by the second Chinese bomb (14 May, 1965). As the 1964–1965 values of ^{134}Cs deposition are probably affected by the contribution of the Chinese bombs, such values have not been taken into account for the determination of the mean residence half-time of the stratospheric ^{134}Cs produced in 1961–1962 tests. On the other hand, the 1962–1963–1966–1967–1968 points seem to indicate a mean residence half-time comparable with that of ^{137}Cs ($\simeq$ 1 yr).[16] On account of the statistical errors affecting the ^{134}Cs measurements, owing to the very low activities involved, an evaluation of the contribution of the Chinese bombs seems to be of little significance.

The amount of ^{134}Cs produced by the 1961–1962 tests was estimated under the assumption of a proportionality between the total deposition at Rome and the total amount of radionuclide produced:

$$\frac{^{137}\text{Cs Total Deposition } (\mu\text{Ci/km}^2)}{^{134}\text{Cs Total Deposition } (\mu\text{Ci/km}^2)} = \frac{\text{Total } ^{137}\text{Cs produced (Ci)}}{\text{Total } ^{134}\text{Cs produced (Ci)}} \quad (1)$$

The ^{137}Cs total deposition was calculated as the integral of the ^{137}Cs monthly deposition[17] corrected for physical decay, over the period

September 1961–December 1965 (95,210 ± 4700 μCi/km^2). The ^{134}Cs total deposition was calculated as the integral over the period September 1961–December 1965, of ^{134}Cs deposition corrected for physical decay with a half-life of 2.1 yr. It is assumed, on the basis of the data for 1961 before the resumption of nuclear tests, that both ^{134}Cs and ^{137}Cs observed in the period under consideration were produced to a great extent, during the 1961–1962 tests. For this reason, the experimental values of 1961 and 1962 depositions were not corrected for physical decay, while the other ones were referred to the middle of 1962. To calculate the integral of ^{134}Cs deposition, the experimental values of the deposition for the period September 1961–December 1963 were used, while the values for 1964–1965 were obtained by extrapolation of the 1963 value under the assumption of a mean residence half-time of 12 months (Fig. 2). The value of the total deposition thus calculated was 150 ± 50 μCi/km^2. The total amount of ^{137}Cs produced in 1961–1962 tests ($\approx$ 100 MT)[19] was assumed to be 1.7 × 10^7 Ci, according to Ref. 1.

A calculation performed according to equation (1) gave a total amount of ^{134}Cs produced in 1961–1962 tests of 27 ± 9 kCi. This value is in good agreement with the estimate (50 kCi) made by Persson.[1] The new injection of ^{134}Cs in December 1968 has to be attributed to the Project Schooner explosion at the Nevada Test Site on 8 December 1968.[18] In fact, ^{134}Cs was one of the radionuclides detected by the U.S.P.H.S. survey network[18] in the period following this explosion. The appearance of an important amount of ^{181}W (one of the predominant long-lived radionuclides produced by the Project Schooner event)[18] was detected, together with a sharp increase of fission products, in air and rainwater samples collected in December 1968 at Casaccia Nuclear Center, Italy.[17] Therefore, it is reasonable to suppose that ^{134}Cs measured in December 1968 fallout (see Table 1), is produced by the above reported event. It must be pointed that ^{134}Cs was undetectable in the fallout monthly samples collected just before December 1968.

CONCLUSIONS

Experimental evidence of ^{134}Cs production exists for the following explosions:

—1955–1956 tests series (5.6 MT,[19]),[10]
—24 June 1957—Priscilla nuclear test (37 ktons, Nevada.[19]),[20,21]
—1961–1962 tests series (100 MT.[19])[1]
—(Not certain) 16 October 1964 First Chinese nuclear test (20 ktons, Lop-Nor,[22]).
—8 December 1968—Schooner nuclear test ($\geqslant$ 35 ktons, Nevada[18]).

As far as the ^{134}Cs production mechanism is concerned, it can be said that no experimental evidence of a production by direct fission is at present available. In fact no correlation can be found between the total yield of the different tests and the corresponding total ^{134}Cs produced. In particular the amounts of ^{134}Cs produced by the 1955–1956 tests, by the first Chinese bomb and by the Schooner event are too great to be accounted for by the hypothesis of production by direct fission only.

Therefore, it is possible to assume, as was pointed out in previous literature,[1,20] that activation of soil plays an important role in the mechanism of ^{134}Cs production at least in the case of explosions occurring near the ground level. Experimental evidence of such a process is given by the presence of considerable amounts of ^{134}Cs in samples of soil collected near the ground-zero point of the Priscilla nuclear test, while ^{137}Cs was practically absent in the same samples.[20,21] As a consequence, it is likely that an activation of stable Cs in soil occurred for the first Chinese bomb (surface experiment) and the Schooner event (cratering experiment).

As far as the great series of nuclear tests are considered, no data on ^{134}Cs production by the 1957–1958 tests are available, while it is difficult to draw any conclusion on the basis of the data referring to the 1961–1962 tests: the great number of powerful explosions in this period could have produced by direct fission an amount of ^{134}Cs ($\approx$ 10 kCi,)[1] comparable with the total estimated ($\approx$ 30 kCi, this paper; $\geqslant$ 50 kCi)[1]. Nevertheless it is not unlikely that a few bombs of this series may have produced, by soil activation, the major amount of ^{134}Cs contamination.

As a conclusion, it seems reasonable to suppose, on the basis of the available data, that the mechanism responsible for the massive production of the ^{134}Cs observed as a world-wide contamination is the activation of stable Cs in the environment. Such an activation, obviously, is dependent on the particular condition in which the explosion occurs more than on its power.

Acknowledgements—The authors take great pleasure to thank Dr. G. Bagliano and Mr. A. Perini for the chemical separation of Cs in the samples.

REFERENCES

1. R. B. R. Persson, *Health Phys.* **14,** 241 (1968).
2. R. W. Perkins, J. M. Nielsen and C. W. Thomas, *Science* **146,** 762 (1964).
3. R. W. Perkins and J. M. Nielsen, *Nature* **205,** 866 (1965).
4. R. W. Perkins, C. W. Thomas and J. M. Nielsen, *Radioactive Fallout from Nuclear Weapons Tests* (Edited by A. W. Klement, Jr.), USAEC, 198 (1965).
5. R. W. Perkins, BNWL, 235/2, 121 (1966).
6. C. W. Thomas and N. A. Wogman, BNWL-481/2, 4 (1967).
7. G. F. Clemente, *Radioactive Dating and Methods of Low-level Counting,* IAEA, 697 (1967).
8. G. Bagliano, M. M. Ferraris and A. Perini, to be published.
9. C. W. Thomas, BNWL 235/2, 117 (1966).
10. G. S. Bien and D. C. Mart, Jr., USAEC, AECU 4164, 13, (Edited by H. E. Suess) (1958).
11. C. M. Lederer, J. M. Hollander and I. Perlman, *Table of Isotopes* (6th ed.); J. Wiley (1967).
12. H. E. Palmer and R. W. Perkins, *Science* **142,** 66 (1963).
13. K. Liden and I. O. Andersson, *Radioactivity in Man* (Edited by G. R. Meneely and S. M. Linde) (2nd Symp.), 488, Thomas. Publ. (1965).
14. W. C. Hanson, D. G. Watson and R. W. Perkins, *Radioaecological Concentration Processes* (Edited by B. Aberg, F. P. Hungate), 233, Pergamon Press, Oxford (1967).
15. L. Cigna Rossi, *Giorn. Fis. San. Radioprot.* **14,** (in press) (1971).
16. C. W. Thomas, J. A. Young, N. A. Wogman and R. W. Perkins, BNWL-SA-1739 (1968).
17. Data to be published.
18. *Rad. Hlth Data Rep.* **10,** 171 (1969).
19. HASL-142, 218 (1964).
20. H. L. Krieger and D. Groche, *Science* **131,** 40 (1960).
21. H. L. Krieger, *Health Phys.* **11,** 653 (1965).
22. Y. Nishiwaki, *Progress in Nuclear Energy,* Series 12, *Health Phys.* (Edited by A. M. F. Duhamel), **2** (1), 662, Pergamon Press, Oxford (1969).
23. P. F. Gustafson and R. B. Keener, *J. Geophys Res.* **70,** 5323 (1965).
24. Y. Nishiwaki, T. Urai, N. Murayama, T. Mamuro, N. Yamagata, M. Saiki and H. Kawai, *Progress in Nuclear Energy,* Series 12, *Health Phys.* (Edited by A. M. F. Duhamel), **2,** (1), 552, Pergamon Press, Oxford (1969).

Environmental Survey of Uranium Mill Tailings Pile, Mexican Hat, Utah

Robert N. Snelling

At the request of the Director, Division of Indian Health, the Bureau of Radiological Health conducted a study of the A-Z Minerals Corporation's tailings pile located at Mexican Hat, Utah. It was requested that the study identify any radiation hazards which might exist and recommend methods for their control. In response to this request, the tailings area was surveyed in May 1968.

The mill was originally operated by the Texas-Zinc Minerals Corporation and transferred to A-Z Minerals Corporation in 1963. Uranium production was discontinued in February 1965. The mill property is located approximately 1 mile southwest of the community of Mexican Hat, Utah (figure 1). The site is situated on the Navajo Indian Reservation and is leased to the company by the Navajo Tribal Council.

The San Juan River flows from east to west approximately 1 mile north of the mill property. Gypsum Creek (actually a dry wash) is located approximately ½ mile east of the tailings pile. The surrounding land is unpopulated and at present is not used for either farming or grazing.

The mill property is divided into three main areas: the tailings area, the mill area, and the housing area (figure 2). The tailings area covers approximately 35 acres and contains an estimated 2.2×10^6 tons of tailings material.[2] The radium content, estimated from the average uranium assay of mill feed material (assuming radium in equilibrium with uranium), is approximately 1,000 pCi/g, or 2,000 total curies.[2] The mill area includes the mill buildings and former stock pile areas. It is located adjacent to and west of the tailings area. The housing area is located approximately 0.5 miles southwest of the tailings area and consists of approximately 60 dwellings, a school, and various community buildings.

Sampling procedures

External radiation measurements were obtained using portable Geiger survey meters. Gamma (closed window) readings were obtained at the surface and at the 3-foot level. A grid was established over the area of interest and readings obtained at 240-foot intervals. A total of 212 pairs of readings was recorded.

In addition to the portable instrument readings, thermoluminescent dosimeters (TLD) were placed

---

[2] Division of Raw Materials, USAEC, 1967.

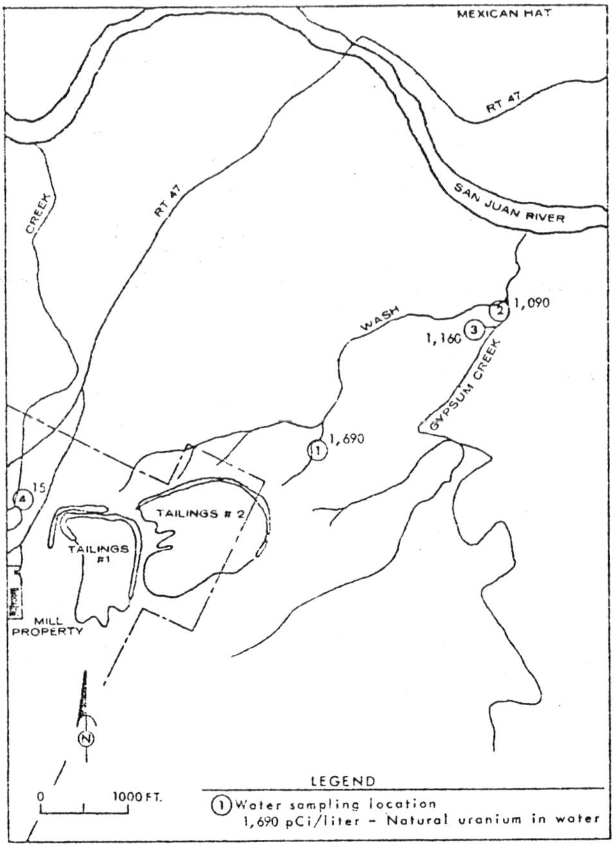

Figure 1. Water sampling locations with natural uranium concentrations Mexican Hat uranium mill, Utah

in triplicate at the 3-foot level at each of the air sampling stations (figure 3). The TLD's respond only to gamma radiation.

Long-lived airborne particulate radioactivity was sampled by drawing air through a 0.8μ pore size filter. A high-volume air sampler was used at a flow rate of about 5-cubic feet per minute. Nine sampling stations were established on and about the tailings area (figure 2). Continuous 24-hour samples were obtained at each station on each of 11 consecutive days. Each individual filter was analyzed for gross alpha radioactivity. In addition, all filters from each station were composited for analysis of natural uranium, radium-226, and thorium-230.

Eight locations were selected for the collection of radon gas samples (figure 3). The sampling technique consisted of pulling filtered air through a small low-volume pump (10 ml/min) into a 30-liter Mylar bag (*1*). A continuous 24-hour sample was collected at each location. The collected samples were picked up by Public Health Service aircraft and flown to Las Vegas for immediate analysis for radon-222.

The short-lived particulate radon daughter products were sampled by pulling air through a glass-fiber filter using a high-volume air sampler. Each sample was collected for 15 minutes at a flow rate of about 10 cfm. Samples were collected at the radon sampling locations (figure 4). The filters were counted for alpha radioactivity immediately in the field utilizing a specially designed alpha scintillation counter. Activity was calculated using the Tsivoglou technique (*2*).

Wind direction and speed were estimated by visual observation. The predominant wind direction at the mill site for the 11-day sampling period was from the southwest. According to A-Z Minerals Corporation personnel, this is the predominant wind direction throughout the year. During the first 7 days of sampling, the wind was light (0–5 mph). During the last 4 days, however, winds estimated at 10–20 mph were encountered during the daylight hours. In general, lapse conditions were encountered during daylight hours with light inversions at night.

Two springs were located in the vicinity of the tailings area at which ground water samples were taken (figure 1). The first was located about 1,000 feet northeast of the tailings area in the wash running from the tailings area (station 1). The wash was dry between the spring and Gypsum Creek. The second spring was just south of the junction of Gypsum Creek and the wash from the tailings area (stations 2 and 3). Gypsum Creek was dry upstream from the spring. In addition to the ground water samples, the drainage from the acid blowdown operation at the mill was sampled (station 4). One-gallon grab samples were obtained at each location, some stations being sampled on several occasions. The samples were analyzed for radium-226 and natural uranium activity.

Samples of tailings material were collected at the locations indicated in figure 4. Approximately 400 cm^3 of surface material was taken from each location. The samples were analyzed for thorium-230, radium-226, and natural uranium.

All laboratory analyses were performed at the Southwestern Radiological Health Laboratory. After sample preparation, gross alpha and gross beta radioactivity determinations were performed by counting directly in an end-window gas flow proportional counter. Radon analysis was performed by cryogenic separation and subsequent alpha scintillation counting. Radium-226 was determined after sample preparation by the radon emanation technique. Uranium analysis was accomplished by the fluorometric technique, and thorium-230 analysis was performed by means of solvent extraction and subsequent alpha-particle counting.

Results

Table 1 summarizes the results of the portable instrument survey on each of the unique areas. Figures 5 and 6 indicate the location and magnitude of the individual measurements at the surface level and at 3 feet. The natural background radiation was found to average 0.03 mR/h both at

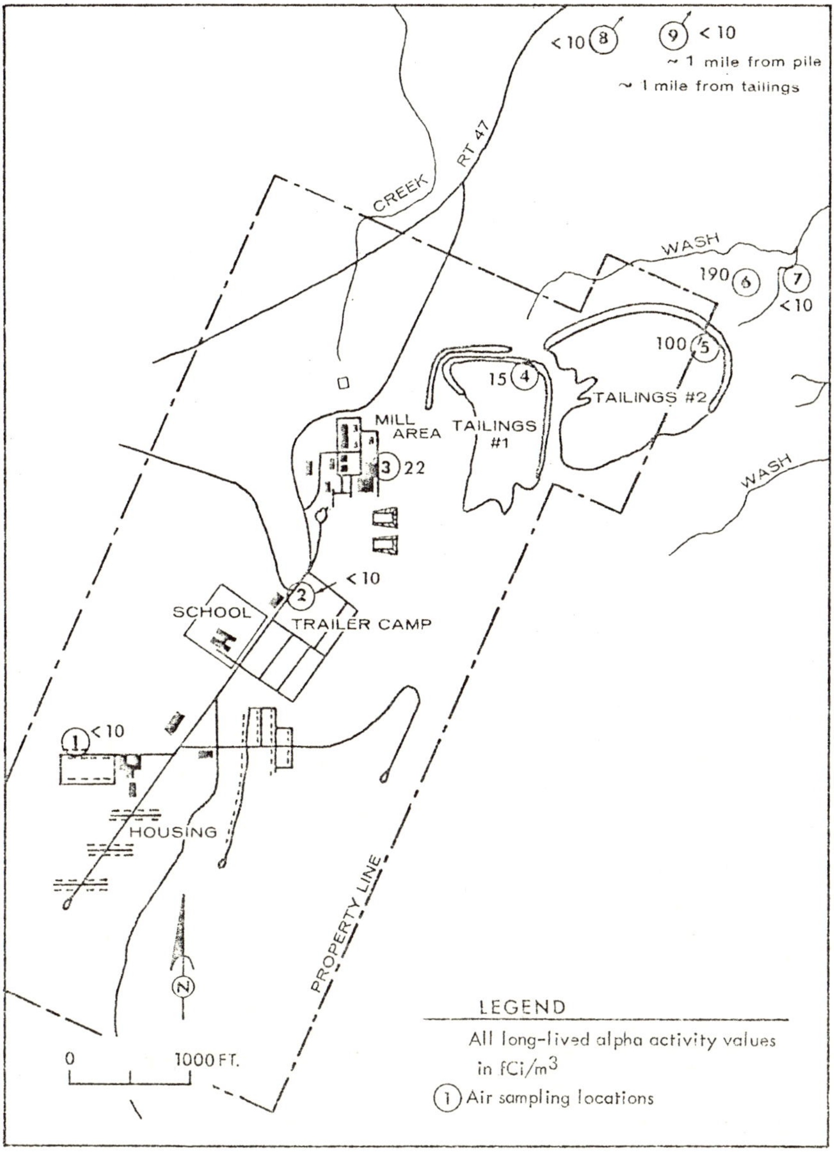

Figure 2. Air sampling locations, Mexican Hat uranium mill

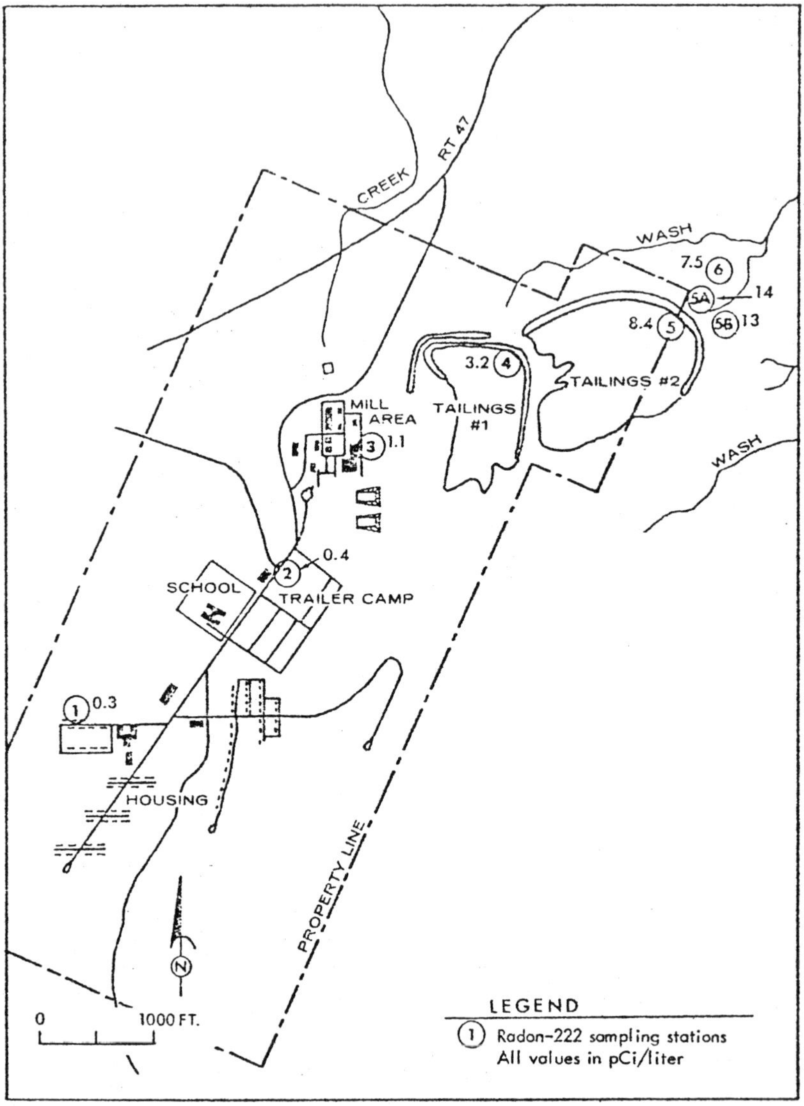

Figure 3. Atmospheric radon-222 sampling stations, Mexican Hat uranium mill

Table 1. External radiation measurements, Mexican Hat uranium mill, May 1968

Area	Number of measurements	Radiation levels at surface (mR/h)		Radiation levels at 3 feet (mR/h)	
		Average	Range	Average	Range
Background	10	0.03	0.02 – 0.06	0.03	0.02 – 0.06
Housing	74	.04	.01 – .2	.03	.01 – .1
Mill	31	.17	.02 – .5	.15	.02 – .5
Tailings	97	.5	.01 – 3.0	.5	.02 – 3.0

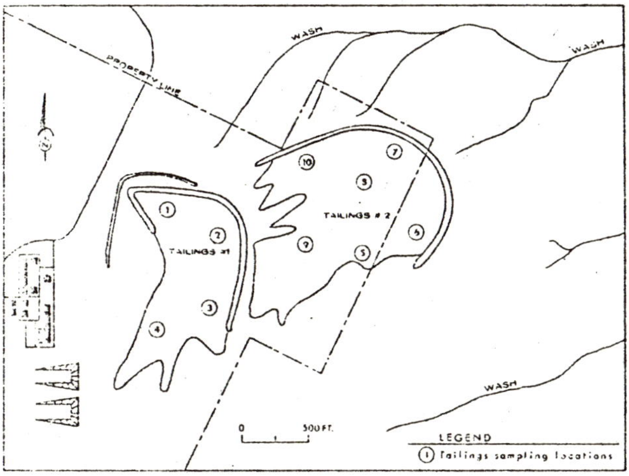

Figure 4. Tailings material sampling locations Mexican Hat uranium mill

the surface and the 3-foot level. The radiation exposure at the 3-foot level on the tailings area ranged from a minimum of 0.02 mR/h on the extremities of the pile to a maximum of 3 mR/h on the pile. The average was 0.5 mR/h. No significant difference was noted between the surface and 3-foot level. The average exposure at the 3-foot level within the mill area was 0.15 mR/h with a range of 0.02 to 0.5 mR/h. Again, there was no significant difference between readings at the surface and at 3 feet. The average exposure at the 3-foot level in the housing area was found to average 0.03 mR/h. This exposure represents back-

Table 2. Comparison of TLD and Geiger meter data May 1968

Station	Exposure at 3 feet (mR/h)	
	TLD[a]	Geiger meter[b]
1	0.01	0.06
2	.03	.05
3	.05	.05
4	.55	.50
5	.42	.40
6	.04	.06
7	.02	.03
8	.01	.02
9	.02	.03
Background 1	.01	.02
Background 2	.01	.03

[a] Eleven day exposure period.
[b] Single survey meter reading.

ground radiation. A comparison of the results obtained from the TLD's and the portable instrument measurements is shown in table 2. The results are in good agreement.

The gross alpha radioactivity found on individual filters is shown in table 3. The background station (station 1) averaged less than 10 fCi/m^3 for 11 days of sampling. Stations 2, 7, 8, and 9 also averaged less than 10 fCi/m^3 indicating background levels of radioactivity. Stations 3 and 4 averaged slightly above background (station 3, 22 fCi/m^3; station 4, 15 fCi/m^3). Stations 5 and 6 were significantly above background levels. Station 5, located on the northeast edge of the tailings pile, averaged 100 fCi/m^3 with a range of 46 to 270 fCi/m^3 for 7 days of sampling. Station 6, located approximately 500 feet northeast from the tailings, averaged 190 fCi/m^3 with a range from less than 10 fCi/m^3 to 510 fCi/m^3 for 11 days of sampling. The average station results are shown in figure 9. It is apparent that the activity concentrations are higher than background in the vicinity of the tailings pile and that the values downwind from the tailings are higher than those upwind.

Table 3. Gross alpha radioactivity in airborne particulates, Mexican Hat uranium mill

Station number	Gross alpha radioactivity (fCi/m³)											
	May 1968				June 1968							
	28	29	30	31	1	2	3	4	5	6	7	Average
1	12	[a] <10	<10	<10	<10	<10	<10	<10	10	<10	<10	<10
2	19	<10	<10	<10	<10	<10	<10	27	<10	<10	50	<10
3	28	11	15	19	14	22	10	57	40	40	18	22
4	NS	<10	9	41	12	<10	15	<10	10	20	61	15
5	270	NS	NS	78	100	46	NS	270	NS	51	66	100
6	<10	15	84	240	36	24	48	360	510	250	370	190
7	<10	<10	<10	24	<10	NS	NS	NS	NS	NS	NS	<10
8	<10	<10	<10	<10	<10	<10	<10	<10	<10	<10	<10	<10
9	<10	<10	<10	<10	<10	<10	<10	<10	12	11	12	<10

[a] Assumed equal to zero for averaging.
NS, no sample.

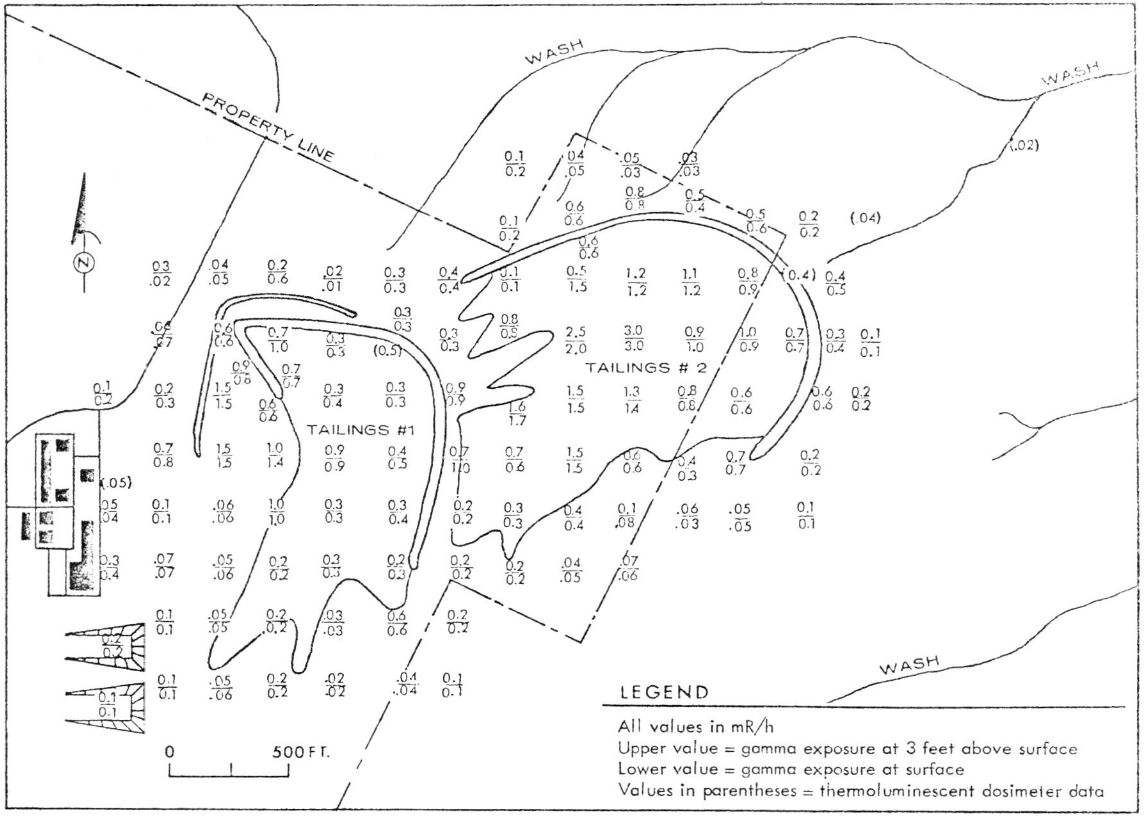

Figure 5. External radiation measurements, Mexican Hat uranium mill tailings piles

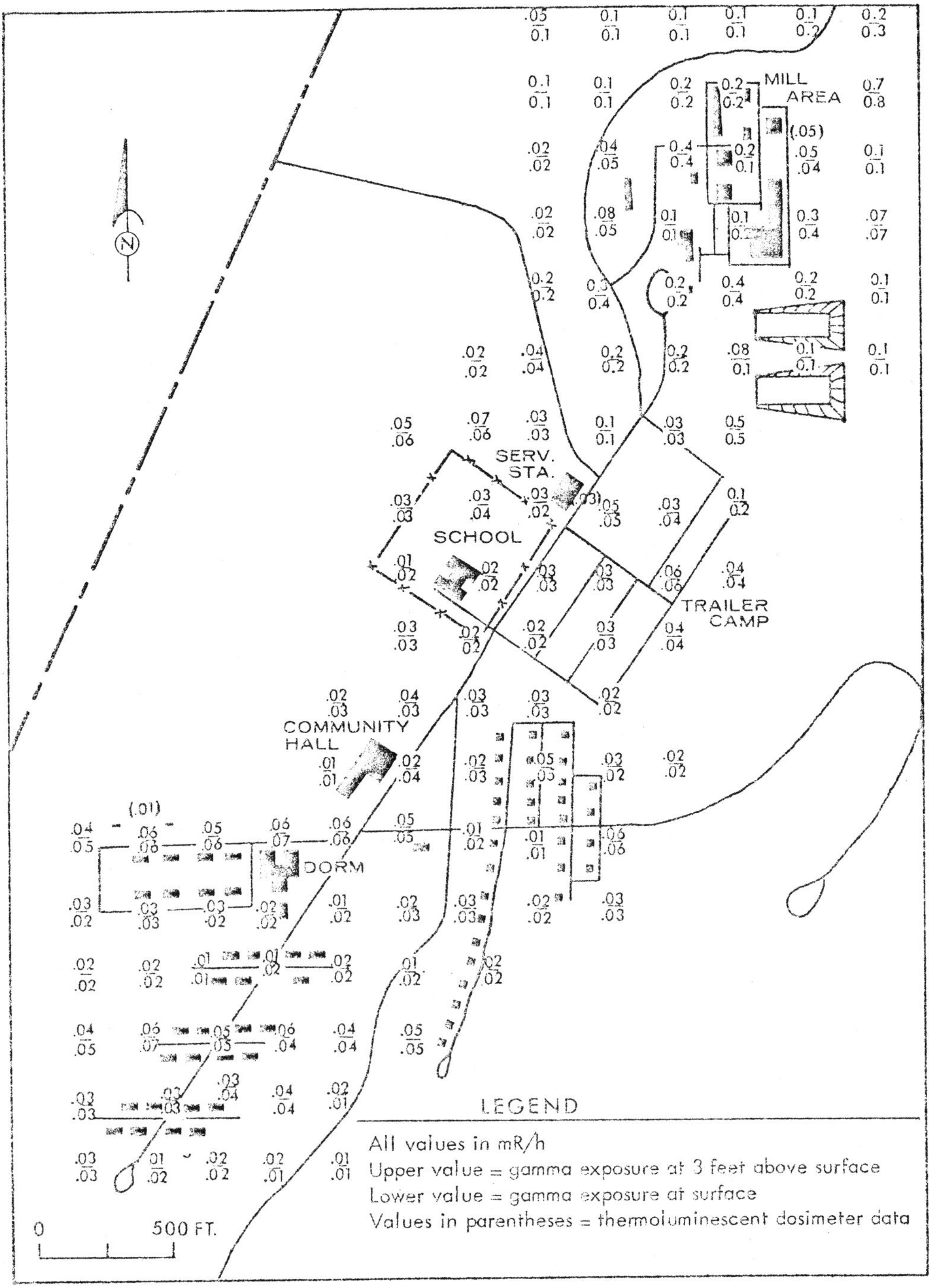

Figure 6. External radiation measurements, Mexican Hat uranium mill

Table 4. Long-lived airborne particulate radioactivity, Mexican Hat uranium mill, May 1968

Station number	Days sampled	Gross alpha radioactivity (fCi/m³)	Radium-226 [a] (fCi/m³)	Thorium-230 [b] (fCi/m³)	Natural uranium [c] (ng/m³)
1	11	<10	0.2	ND	0.2
2	11	<10	.2	ND	1
3	11	22	2	7	10
4	10	15	1	3	2
5	7	100	5	7	3
6	11	190	7	14	4
7	5	<10	.7	ND	.6
8	11	<10	.2	ND	1
9	11	<10	.3	ND	1

[a] Minimum detectable activity for radium-226 — 0.1 pCi/sample.
[b] Minimum detectable activity for thorium-230 — 5.0 pCi/sample.
[c] Minimum detectable activity for natural uranium — 0.1 μg/sample.
ND, nondetectable on composited sample.

Table 4 shows the average gross alpha radioactivity along with radium-226, thorium-230, and natural uranium activity for each station. It should be noted that stations 3–6 (those indicating positive gross alpha radioactivity results) display radium concentrations averaging about a factor of ten higher than the other stations. These same stations show significant concentrations of thorium-230, whereas all other stations do not. Although the natural uranium concentrations average slightly higher for stations 3–6, the increase does not appear to be significant with the exception of station 3.

Table 5 summarizes the radon-222 activity concentrations found at various sampling times and locations. Figure 3 represents the average radon concentrations at each sampling location. Station 1 (a background station) showed a radon concentration of 0.3 pCi/liter. Station 2 (0.4 pCi/liter) can also be considered background level. Those stations on and about the tailings area showed levels of radon significantly above background. Individual samples ranged from a minimum of 0.6 pCi/liter at station 3 to a maximum of 14 pCi/liter at station 5A.

Table 5. Atmospheric radon-222, Mexican Hat uranium mill, May–June 1968

Station number	Midpoint of sampling		Radon-222 (pCi/liter)
	Date	Hour	
1	5/27	2030	0.3
2	5/28	2300	.4
3	5/29	2100	1.6
3	6/6	0030	.6
4	5/30	2330	5.7
4	6/4	2400	.7
5	5/31	2100	8.4
5A	6/3	2300	14.0
5B	6/2	2300	13.0
6	6/1	1930	7.5

Table 6. Particulate radon daughter products in air, Mexican Hat uranium mill, June 1968

Station number	Collection		Concentration (pCi/m^3)		
	Date	Time	Polonium-218	Lead-214	Bismuth-214
5B	6/2	2000	120	90	80
5B	6/3	1100	210	50	10
5A	6/3	1400	10	30	30
5	6/3	1900	0	10	100
5	6/4	1045	0	120	160
4	6/4	1430	0	3	60
3	6/5	1400	220	40	0

Table 7. Long-lived radioactivity in water, Mexican Hat uranium mill, May 1968

Station number	Location	Number of samples	Radium-226 (pCi/liter)	Natural uranium[a] (μg/liter)
1	Wash (spring)	3	1.5	1,690
2	Gypsum Creek	2	.2	1,090
3	Gypsum Creek (spring)	1	.2	1,160
4	Acid blowdown	2	.5	15
Background	Monument Valley	5	.5	2

[a] 1 μg/liter of natural uranium equals 0.7 pCi/liter of alpha radioactivity.

Table 6 summarizes the results of the radon progeny evaluation. It should be noted that the levels of activity are several orders of magnitude lower than the radon concentrations given in table 5. The apparent disequilibrium between polonium-218 and its daughter products may be due in part to poor counting statistics at the low levels of activity counted. It must be assumed, therefore, that a very low state of equilibrium existed between the radon and its progeny during the survey period.

Table 7 shows the average radium-226 and natural uranium in water results for each sampling location. Because there were no suitable ground water sources in the immediate area, ground water samples from the Monument Valley area (approximately 22 miles distant) are used as background indicators. The uranium levels in the ground water samples were significantly above background, averaging between 1,090 μg/liter and 1,690 μg/liter. The acid blowdown water showed a natural uranium concentration of 15 μg/liter which is only slightly above background. Figure 1 shows the average natural uranium concentration for each location.

Table 8 shows the results of the thorium-230, radium-226, and natural uranium analysis on tailings material. The radium content averaged 370 pCi/g. This average is somewhat below that predicted from the average assay of mill feed material (1,000 pCi/g). Thorium-230 averaged 1,960 pCi/g with individual values ranging from a minimum 220 pCi/g to a maximum of 3,700 pCi/g.

Discussion

The Code of Federal Regulations, Title 10, Part 20 (10 CFR 20), indicates a maximum allowable whole body dose to an individual in an unrestricted

Table 8. Tailings analysis, Mexican Hat uranium mill, May 1968

Area	Number	Uranium ($\mu g/g$)	Thorium-230 (pCi/g)	Radium-226 (pCi/g)
Tailings #1	1	63	2,400	340
	2	350	3,000	130
	3	69	3,700	31
	4	27	3,200	27
Tailings #2	5	40	2,800	860
	6	70	220	580
	7	73	220	390
	8	470	2,400	550
	9	120	480	320
	10	180	1,200	470
Average		150	1,960	370

area of 0.5 rem in any period of 1 calendar year. The regulations further indicate a maximum allowable dose for continuous exposure to individuals in an unrestricted area of 2 millirems in any 1 hour or 100 millirems in 7 consecutive days.

The Radiation Protection Guide (RPG) adopted by the Federal Radiation Council recommends that the yearly radiation exposure to the whole body of individuals in the population (exclusive of natural background and the deliberate exposure of patients by practitioners of the healing arts) should not exceed 0.5 rem (*3*). The Council further recommends, however, that every reasonable effort should be made to keep exposures as far below this level as possible.

To evaluate the potential long-range hazard associated with the external radiation emitted from the tailings material, the average yearly whole-body exposure for an individual having 24-hour-per-day residence on the area was calculated and compared to the recommended RPG of 0.5 rem/a. The net gamma radiation (i.e., minus background) at 3 feet was used for this computation.

The average exposure at the 3-foot level on the tailings area was found to be 0.5 mR/h (background was assumed negligible). The whole-body average yearly dose for an individual having 24-hour-per-day residence on the tailings area, on the basis of this exposure rate, is approximately 4.4 rem/a. This is nine times the recommended RPG of 0.5 rem/a. These findings serve to preclude the release of the tailings area for public use. Action which would permit release of the area would involve covering the tailings with uncontaminated soil to an extent that would diminish the external radiation to an acceptable level and to stabilize the covering against wind erosion. Otherwise, the area should be fenced and controlled as a radiation area.

The average radiation exposure at the 3-foot level around the mill area was 0.15 mR/h. Subtracting an average background of 0.03 mR/h yields an average net exposure of 0.12 mR/h. The yearly average dose to an individual having 24-hour-per-day residence in this area, on the basis of the average value of 0.12 mR/h, is 1 rem/a. This dose is twice the RPG of 0.5 rem/a but represents the highly unlikely condition of continuous occupancy. With reasonable occupancy assumptions it would not be expected that a yearly average dose in excess of 0.5 rem would occur.

The radiation exposure at 3 feet in the housing area was found to consist entirely of background radiation.

In order to assess the potential long-range health hazard associated with the inhalation of airborne radionuclides, the average activity concentrations observed for each sampling station are compared to the recommended radiation concentration guides as indicated in 10 CFR 20. Because the guides are applicable to yearly average concentrations, this comparison assumes that the meteorological conditions imposed during the sampling period are representative of the yearly average conditions.

The radiation concentration guide for unidentified natural radionuclides in air, specified in 10 CFR 20, is 80 fCi/m^3 (when protactinium-231 is assumed absent). If gross alpha radioactivity is above this guide, identification of the specific radionuclides contributing to the total alpha radioactivity is necessary in order to evaluate the potential long-range hazard associated with the inhalation of the airborne particles.

The nuclides in the natural uranium decay chain most significant as potential health hazards are the alpha-particle emitters. These nuclides and their appropriate guides are shown below:

Thorium-230 (soluble)	80 fCi/m^3
Radium-226 (insoluble)	2 pCi/m^3
Natural uranium (insoluble)	2 pCi/m^3
Lead-210 (soluble)	4 pCi/m^3
Polonium-210 (insoluble)	7 pCi/m^3

It is apparent from comparison of these values with table 4 that none of those stations showing elevated gross alpha concentrations exceeds the guides for individual radionuclides. The following conclusions can be made; the tailings pile is resulting in elevated concentrations of radioactivity in air in the downwind direction (at the present time, however, these concentrations cannot be shown to exceed the recommended guides for individual isotopes), and it is impossible to predict whether the condition will be worse in the future. At present, the pile is still quite moist and displays a clay-like consistency. As the pile dries out, the condition could very well become worse. On this basis two courses of action are possible. The tailings area may be stabilized against wind erosion. This action would preclude further carriage of tailings material into the surrounding environment in the future and would eliminate the potential for any long-term hazard to anyone inhabiting the area downwind from the mill property. If the tailings area is not stabilized, periodic monitoring will be necessary to insure that the recommended concentration guides are not exceeded.

The Code of Federal Regulations, Title 10, Part 20, indicates a radiation concentration guide for radon-222 in air in unrestricted areas of 3 pCi/liter above natural background. The daughters of radon-222 are assumed to be present to the extent that they exist in unfiltered air (this is interpreted as meaning equilibrium conditions). To evaluate the potential long-range hazard associated with the inhalation of radon and its progeny, the atmospheric radon concentrations are compared to this guide. Average radon concentrations in the tailings area ranged from a minimum of 3.2 pCi/liter at station 4 to a maximum of 14 pCi/liter at station 5A. These values are equal to or greater than the concentration guide of 3 pCi/liter. It was found, however, that a very low percent equilibrium existed between radon and radon daughters. Since approximately 95 percent of the dose to the lungs due to inhalation of radon and its daughters is due to the particulate daughters, the elevated radon levels cannot be considered as representing violation of the guide. It should be pointed out, however, that if enclosed structures were to be built on the tailings material a substantial increase in both radon and daughters could be expected within the structure; and the radiation concentration guide may be exceeded. Therefore, it is recommended that no enclosed structures be built on the tailings material. Station 3 indicated an average radon concentration of 1.1 pCi/liter. Subtracting natural background gives a net average concentration of 0.8 pCi/liter. This is well below the appropriate concentration guide. The housing area (stations 1, 2) showed essentially background levels of radon-222.

Ground water samples obtained in the vicinity of the tailings area showed natural uranium concentrations in excess of 1 nCi/liter. These values are significantly above background. The Code of Federal Regulations, Title 10, Part 20, indicates a radiation concentration guide for natural uranium of 20 nCi/liter. It is apparent that the concentrations observed are a factor of ten below the recommended guide.

The concentration guide for soluble radium-226 is 30 pCi/liter. None of the samples collected exceeded this guide. It must be concluded. therefore, that although ground water samples in the vicinity of the tailings showed uranium concentrations significantly above background, they are still well below the recommended concentration guides.

Summary

External gamma radiation on the tailings area averaged 0.5 mR/h at 3 feet above the surface. By calculation this comes to a yearly dose of 4.4 rem/a for continuous exposure and exceeds by a factor of nine the recommended Radiation Protection Guide (RPG) of 0.5 rem/a for continuous whole body exposure to individuals in the general population. These findings serve to preclude the release of the tailings area for public use. Action which would permit release of said area would be to cover the tailings with uncontaminated soil to an extent that would diminish the external radiation to an acceptable level and to stabilize the covering against wind erosion. Otherwise, the area should be fenced and controlled as a radiation area.

The tailings pile extends approximately 300 to 500 feet beyond the present property line in the easterly direction. The size of the pile precludes the possibility of returning this material to the property area and therefore this area should be fenced and controlled as a radiation area.

The mill area exhibited external gamma radiation levels averaging 0.12 mR/h above background at 3 feet above the surface. By calculation this comes to a yearly average dose of 1.0 rem for continuous exposure which is two times the appropriate RPG of 0.5 rem/a. With reasonable occupancy assumptions, however, the yearly average dose would probably not exceed 0.5 rem.

The housing area displayed background levels of external radiation.

Airborne transport of tailings material in the northeast direction from the pile is demonstrated by higher than background levels of radioactivity in air. None of the stations showing elevated gross alpha concentrations exceeded the concentration guides for radium-226, thorium-230, or natural uranium, however. Since it is impossible to predict whether the condition will become worse in the future with the drying out of the tailings pile, it is recommended that the area be stabilized against wind erosion. If the area is not stabilized, periodic monitoring will be necessary to insure that the appropriate concentration guides are not exceeded.

The gross alpha radioactivity, radium-226, thorium-230, and natural uranium concentrations in the mill and housing areas were well below the applicable radiation guides.

Average radon gas concentrations on and about the tailings area ranged from 3.2 pCi/liter to 14 pCi/liter. These levels exceed the concentration guide of 3 pCi/liter. It should be pointed out, however, that this guide is applicable to radon-222 and its daughters "as they occur in unfiltered air." This may be interpreted as meaning radon in equilibrium with its daughter products in air. Since it was demonstrated that equilibrium did not exist between radon and its progeny, at least at the time of daughter product sampling, it does not appear that the concentration guide has been exceeded. Levels of radon-222 in the mill area and housing area are substantially below the recommended guide.

It should be emphasized that if enclosed structures were to be built on the tailings material, a substantial increase in radon concentration would occur within the structure and the concentration guide may be exceeded.

Ground water samples obtained in the vicinity of the tailings area showed natural uranium concentrations in excess of 1 nCi/liter. These values represent concentrations substantially above background but are well below the appropriate concentration guide.

Recommendations

On the basis of the data gathered in this survey, it is recommended that the mill tailings be properly stabilized against wind erosion. This action would preclude further transport of tailings material into the surrounding environment and would minimize the potential long-term hazard to anyone inhabiting the area downwind from the mill property. If the tailings area is not stabilized, periodic surveillance will be necessary to insure that significant wind carriage does not occur.

As a result of the external radiation levels on the tailings area itself, this area should not be released for public use in its present state. Action which would permit release of the area would be to cover the tailings with uncontaminated soil to an extent that would diminish the external radiation to an acceptable level and to stabilize the covering against wind erosion.

Acknowledgments

Sincere appreciation is expressed to the A-Z Minerals Corporation for the cooperation and assistance provided during the survey.

REFERENCES

(*1*) SILL, C. An integrating air sampler for determination of ^{222}Rn. Health Physics 16: 371–377 (March 1969).

(*2*) TSIVOGLOU, E. C., H. E. AYER, and D. A. HOLADAY. Occurrence of nonequilibrium atmospheric mixtures of radon and its daughters. Nucleonics 11 (September 1953).

(*3*) FEDERAL RADIATION COUNCIL. Background material for the development of radiation protection standards, Report No. 1. Superintendent of Documents, U.S. Government Printing Office, Washington, D.C. 20402 (May 13, 1960).

AUTHOR INDEX

KEY-WORD TITLE INDEX